Manufacturing Competency and Strategic Success in the Automobile Industry

Manufacturing Competency and Strategic Success in the Automobile Industry

By
Dr. Chandan Deep Singh and
Dr. Jaimal Singh Khamba

CRC Press
Taylor & Francis Group
Boca Raton London New York

CRC Press is an imprint of the
Taylor & Francis Group, an **informa** business

CRC Press
Taylor & Francis Group
52 Vanderbilt Avenue,
New York, NY 10017

First issued in paperback 2020

© 2019 by Taylor & Francis Group, LLC
CRC Press is an imprint of Taylor & Francis Group, an Informa business

No claim to original U.S. Government works

ISBN-13: 978-1-138-59851-5 (hbk)
ISBN-13: 978-0-367-65668-3 (pbk)

Library of Congress Cataloging-in-Publication Data

Names: Singh, Chandan Deep, author. | Khamba, Jaimal Singh, author.
Title: Manufacturing competency and strategic success in the automobile industry / Chandan Deep Singh and Jaimal Singh Khamba.
Description: Boca Raton : Taylor & Francis, a CRC title, part of the Taylor & Francis imprint, a member of the Taylor & Francis Group, the academic division of T&F Informa, plc, 2019. | Includes bibliographical references and index.
Identifiers: LCCN 2018041667| ISBN 9781138598515 (hardback : acid-free paper) | ISBN 9780429486302 (ebook)
Subjects: LCSH: Automobile industry and trade–Management. | Soft skills. | Success in business. | Automobile industry and trade–India–Personnel management–Case studies. | Automobiles–India–Design and construction–Case studies.
Classification: LCC HD9710.A2 S57 2019 | DDC 338.4/76292220684–dc23
LC record available at https://lccn.loc.gov/2018041667

Visit the Taylor & Francis Web site at
http://www.taylorandfrancis.com

and the CRC Press Web site at
http://www.crcpress.com

Contents

Preface

The strategic success of an industry depends upon manufacturing competencies, that is, their competitive advantage. If the industry has the parameters of better quality and reliability, this will lead to increased sales and the creation of a sound customer base for greater market share, thus returning with more profit, growth, and expansion. Competitive priorities are the operating advantages that a firm's processes must possess to outperform its competitors. The operating advantage for the industry is assessed, evaluated, and measured with the parameters of cost, quality, time, design, flexibility, etc. This book is so designed to surpass the expectations of industrialists, policymakers, and competency designers; specifically, the manufacturing competencies upon which the whole strategic success of the industry depends. Quality, cost, delivery, innovation, and responsiveness influence most manufacturing strategic agendas today. Firms have traditionally pursued these goals through the adoption of advanced technologies and manufacturing practices, such as concurrent engineering, JIT, and worker empowerment. Recent developments in the industry suggest the emergence of another route to manufacturing excellence, that is, there should be an increasing focus by industry regulators and professional bodies based on the need to stimulate innovation in a broad range of manufacturing competencies. By 'competencies' we mean the methods, equipment, and expertise that can be developed as a leading capability in one market sector or application, and have potential to be applied successfully across other sectors or applications as well. Further, competencies are the ability to apply or use a set of related knowledge, skills, and abilities to perform 'critical work functions' or tasks in a defined work setting. Competencies often serve as the basis for skill standards that specify the level of knowledge, skills, and abilities required for success in the workplace, as well as potential measurement criteria for assessing competency attainment. Strategies are actions a business takes to compete more aggressively, to acquire additional customers, and to operate the company more profitably. A successful strategic plan provides the information and guidance that the management team needs to run the company with greater efficiency and help the business reach its full potential. Strategic planning helps managers make decisions based on logical assumptions and a clearer view of the future. The strategic success of the industry is related to profitability, market share, growth and expansion, quality and reliability, labour intensiveness, etc. For accomplishing a success set of parameters, the operations strategy links long- and short-term operations decisions to corporate strategy, which is composed of core competencies – these are the unique resources and strengths of the organization, which include workforce, facilities, market and financial know-how, and systems and technology.

About the Authors

Chandan Deep Singh has been serving as an assistant professor in the Department of Mechanical Engineering at Punjabi University, Patiala, Punjab, India since 2011. He completed his PhD in November 2016 from the same institution. His masters of technology in manufacturing systems engineering is from the Sant Longowal Institute of Engineering and Technology in Longowal, Sangrur, Punjab, India, completed in 2011. He completed his bachelor of technology in mechanical engineering in 2009 from Giani Zail Singh College of Engineering and Technology in Bathinda, Punjab, India. He has published around 52 books and guided 55 students for their master of technology thesis. He has published around 98 papers in various international journals and conferences. Presently, 7 students are working under him for their PhD and 1 for his masters of technology. His main research areas are CAD/CAM, production and industrial engineering, and die-casting. He has worked on software, namely CATIA, ProE, Solid Works, PSAW, MS-Excel (for AHP, TOPSIS, VIKOR), AMOS (in PSAW for SEM), and MATLAB.

Jaimal Singh Khamba holds a bachelor in mechanical engineering, master in industrial engineering, and PhD in technology management from the Thapar Institute of Engineering and Technology, Patiala, Punjab, India. He is currently a professor of mechanical engineering at Punjabi University, Patiala, Punjab, India. He has more than 200 publications in refereed journals and conferences. He has guided 11 Ph.D. students including Chandan Deep Singh and 5 students are working for their Ph.D. under his guidance. His main research areas are Non Traditional Machining (Ultra Sonic Machining), TPM, and manufacturing competency.

1

Competency and Its Components

The Indian automobile industry has been witnessing the entry of global automobile giants like Volkswagen, Mercedes, and Audi since the beginning of twenty-first century, leading to fierce competition for already existing players, like Japanese automaker Suzuki, Korean automaker Hyundai, Italian automaker Fiat, and such others. This has led to a significant increase in competition for the survival of automobile manufacturers in India. For better survival in any market, companies need to be agile, that is, they have to produce innovative products quicker than their competitors. Innovation is directly related to designing and producing new products or making a few desirable changes in the existing ones to satisfy ever-demanding customers.

1.1. Competency

Competency is the collective application of knowledge, skills, and behavior for enhancing organizational achievements, and helping the organization realize specific goals and objectives. Management competency encompasses emotional intelligence, systems thinking, and skills in negotiation and influence. Competency attributes directly enhance the functional performance that has considerable impact on overall organizational staff and functions.

Competencies are progressive in nature and offer direct benefits to organization and personnel, they enrich employee accomplishments and attributes, and help management create proactive transformation in organizational culture for facing global competition and future challenges. Competencies facilitate organizations in the framing of goals and objectives, especially within human resources (HR). They also provide a framework for objective proficiency and consistent standards by creating shared language about well-depicted organizational specifications and requirements. Technical competencies include proficiencies and skills related to processes, roles, and functions within the organization, which cater to the development and applications of related procedures, policies, and regulations relevant to the particular business or technical field.

Competency has also been characterized as catering to HR require-ments in organizations and communities. Competency is defined as learning from prior context and situations that might be different the next time a person has to act. During crunch situations, competent managers will tackle a particular situation by adopting similar solutions from previous successful experience. Thus, competent managers must interpret situations in context and should develop an appropriate repo-sitory of possible solutions through training. Competency matures over a period of time through experience and knowledge irrespective of training.

Competencies yield enhanced performance levels by fostering and enriching appropriate skills, knowledge, and abilities by individuals or organizations, and provide a framework for distinguishing between poor and exceptional performance. Competencies may have organizational, team, or individual attributes that add significantly to manufacturing performance enhancement. These divergent perspectives suggest that an alternative approach might be useful to stay ahead of the competition (Hoskisson et al., 1999).

Different aspects of competency are depicted in Figure 1.1. Competency acquisition by an organization's personnel induces enhanced skills and abilities that facilitate the accruing of improved organizational perfor-mance through the clarification of job requirements. Competencies allow for opportunities for improvement in existing job profiles.

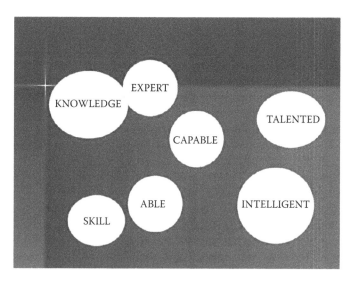

FIGURE 1.1
Different Aspects of Competency.

Competencies can be evolved with individual and organizational contributions and endeavors. Top management can identify and manage competencies that facilitate work procedures, effectiveness, and enhancement in human skills and competencies, which can be integrated with organizational learning involving on-the-job (OJT) experience, classroom learning, or other training opportunities.

1.1.1. Why Competencies?

Since the global business competition has demonstrated a transition to innovativeness, efficiency, and value addition from economies of scale, management should strategically focus on harnessing employee-centric competencies. Strategy is the direction and scope of the organization; ideally, it synchronizes with its assets to its changing competencies. Figure 1.2 shows different levels in competence study.

Therefore, effective evolution and management of HR competencies assume significant importance for the organization's pursuit of excellence and survival. Thus, organizations need to foster HR management (HRM) competency systems that effectively apprehend wide variety of skills of personnel, encourage multiple job attributes, and permit flexibility in incentive decisions for catering to dynamic organizational attributes. The competency development domain is continuously gaining administrative management acceptance among business organizations globally. Competency models have the potential for facilitating organizations to finalize important business decisions.

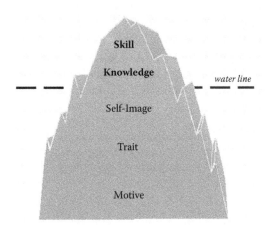

FIGURE 1.2
Levels in Competence Study (www.astd.org).

The following issues highlight the need for fostering competencies:

1. For apprehending performance, it is pertinent to observe the traits and pursuit of successful personnel, rather than following a set of assumptions pertaining to traits and intelligence
2. Competencies are the tools to evaluate personnel performance at the workplace
3. Competencies can be mastered and harnessed
4. Competencies should be highlighted and made accessible
5. Competencies should be correlated to meaningful process end results depicting personnel performance at the workplace

Core competencies are not specific to any particular business or organization. It has been observed that teamwork, participative management, and customer focus are crucial competencies that apply equally well in different business domains. Further, certain specific technical competencies might not be appropriate under various industrial domains. Moreover, some departments can evolve function-specific competencies to complement the core competencies appropriate to suit their specific work requirements. Competency may require skill sets related to a particular field of technology or skill. The capabilities or skill sets should be well maintained and practiced at optimum levels as much as possible.

Invariably, competent personnel must be deployed in specific areas of the workplace. These personnel can exhibit their capabilities through efficient workplace training, formal or informal qualifications, and sharing their knowledge gained over a period of time. Further, specific teams can be appropriately deployed for developing and demonstrating competence. The personnel should evolve, adopt, and exhibit individual competence levels for the realization of organizational objectives. Competence development involves an organization's endeavor to evolve their existing competence status, including both traditional teaching-learning attributes, as well as practical OJT methods. Organizational market performance is the firm's market performance evaluation compared to overall industry-wide performance.

> An organization's competitiveness involves a set of distinct technological attributes, complementary assets, and organizational practices leading the organization's competitive capabilities in one or more businesses.

Thus, personnel with higher experience and skill attributes will show superior improvements and produce fewer surprises over less experienced or less prepared ones (Levinthal and March, 1993). Competency is frequently proclaimed as possessing the appropriate skill sets, potential, authority, and qualification. Domain knowledge, aptitude, attitude, commitment, and motivation have been accepted as key attributes for garnering and nurturing effective competencies related to specific tasks or context. Competency is closely associated with workplace performance and should be evaluated in terms of productivity. competence is explained as:

> Collective application of personnel skills, expertise, and aptitude for delivering productivity objectives of an organization.

1.1.2. Classification of Competencies

Competencies can be classified in different ways, and each type of competency has different significance and relevance.

1.1.2.1. Core Competencies

A core competency is defined as an internal capability that is crucial to the prosperity of any organization. Personnel should possess these basic core competencies in order to realize successful organizational performance and in turn dictate organization values (Bonjour and Micaelli, 2010). Some of the personnel core competencies traits can be teamwork, motivation, flexibility, aptitude, and interpersonal skills. Personal core competencies demand individuals to be able to perform effectively in diverse applications at desired performance levels anywhere in the organization.

Core competencies may vary depending upon the nature of job or technology used in a particular organization. For example, whereas an electronic equipment manufacturer may require specialization in the design of electronic components and circuits, the software organization might call focus on key skills involving high quality software code writing. Core competencies should evolve continuously with a rapidly changing organizational environment. Therefore, core competencies must be flexible and evolve over the time a business/industry progresses, leading to growth and opportunities.

Prahalad and Hamel (1994) described core competencies as collective learning in a firm, especially the integration of multiple streams of technologies and coordinate diverse production skills. In the short term, an organization's competitiveness derives from the price or performance attributes of products, whereas in the long term, it derives from the ability to develop, more quickly and at a lower cost than competitors. They described 'competitor differentiation, extendibility and customer value' as the conditions for core competency.

With core competency, an organization should be able to provide benefit to customers as well as expand into new markets. Also, competitors should find it hard to replicate. Core competency thinking promotes approach to mobilize and focus on an organization's resources. Technology executives and research and development should describe the core competencies of their companies, as core competency thinking enables competitive advantage without disruption to business activities (Gallon et al., 1999). Core competency has a dynamic role in improving the potential of project teams. Core competency improves strategies by balancing itself with the external environment and activities, by reducing path-dependent influences, and by carefully arranging resources by guidance rather than control (Ljungquist, 2007, 2013).

Core competencies affect entrepreneurial performance and business success. Core competencies develop strategies in relation to the firm's performance (Mitchelmore and Rowley, 2010). Core competencies enable firms to develop different perspectives in accordance with company values and strategies (Nyhan, 1998). Core competencies have an important role in Indian organizations in different conditions. SMEs should be as proactive in making changes, such as the development of competencies, awareness about market changes, technology upgrades, and HR (Singh et al., 2008).

Core competencies should enhance strategic thinking in order to achieve the organization's goal. Without the core competencies, well-stated and well-conceptualized strategies cannot be successfully realized and implemented. Core competencies are important for organizations in achieving competitive advantage (Cardy and Selvarajan, 2006). They are linked to business profitability, as they are cross-culturally valid. Core competency models should focus on selection, feedback, training, and performance management (Ryan et al., 2012).

Core competency models focus on competency scoring, competency identification, and aligning competency with strategic functions to survive in the competition. Organizations will manage their employee competencies to ensure a competitive advantage and better performance (Sengupta et al., 2013). Core competencies help individuals to perform better in different positions throughout the organization. Various competency models are used to identify and expand the global competitiveness of organizations.

1.1.2.2. Professional Competencies or Functional Competencies

A professional or functional competency means the capability to perform certain tasks satisfactorily. These competencies are usually job-specific and demand high performance and quality results for a particular task or job. These competencies are usually technical or operational in nature, requiring peculiar skill sets, and are often managed at a workgroup level. They must be appropriately evolved for different roles and functions depending upon the function's maturity. Examples of functional competencies include database management systems, C++ programming, reliability assessment, and security systems.

1.1.2.3. Behavioral Competencies

Behavioral competencies mean measurable and observable personal behaviors. These competencies call upon evaluating knowledge, skills, aptitudes, abilities, and other related personal attributes that are necessary for individual's pursuit for attaining excellence at work. These traits involve the ability of individuals to excel in teamwork, managerial, analytical, and communication skills; as well as confidence, motivation, and interpersonal leadership skills.

1.1.2.4. Threshold Competencies

Threshold competencies mean the bare minimum capabilities to perform a given task satisfactorily and should be possessed by personnel to be able to perform a job effectively. They do not distinguish between an average and superior performer. These may include basic knowledge, traits, self-image, and social roles. For example, a typist must possess primary typing skills and language knowledge which constitute a threshold competency, but this does not guarantee the correctness or appropriateness of the text. Similarly, good marketing personnel should have appropriate knowledge about the products, but this itself does not guarantee outstanding performance.

1.1.2.5. Differentiating Competencies

Differentiating competencies include the attributes that differentiate superior performers from average performers. For example, typists that possess formatting skills have the potential to exceed their performance, thereby adding to their differentiating competency. Similarly, customer-focused and empathetic marketing personnel can appropriately judge the customer's requirements to be able to deliver and sell more, thereby realizing superior performance.

1.1.3. Manufacturing Competencies

Manufacturing competencies are the combination of pooled knowledge and technical capacities that allows a business to stay competitive in the market. Theoretically, it should allow a company to provide significant benefit to customers, as well as expand into new end markets. Also, it should be hard for competitors to replicate. Figure 1.3 depicts a conceptual framework for competencies.

Manufacturing competencies have different perspectives. One perspective is that it involves an understanding of specific phenomena and their

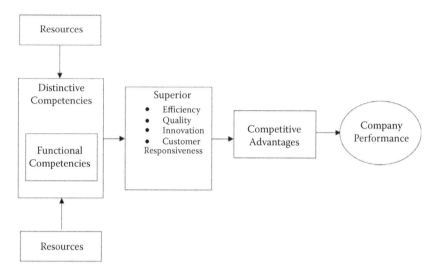

FIGURE 1.3
Conceptual Framework.

related disciplines. Examples of these phenomena can include electronics, pharmaceuticals, etc. Examples of related disciplines include mechanical engineering, biochemistry, physics, etc. A second perspective defines an organizational competency to include technology (such as internal combustion, computing, or printing) and its related products. For example, a computer is an end product, a hard drive is a core product, and the underlying competency lies in the technology of magnetic data storage. A third suggests that an organizational competency usually includes functional skills. Examples include distribution, manufacturing, production scheduling, or marketing. A fourth perspective suggests that an organizational competency includes integration, usually of skills and technology. For example, Maruti Suzuki successfully obtained synergy between internal combustion technology with the engineering and manufacturing functional attributes for producing excellent quality automobile engines (Prahalad and Hamel, 1994).

Competencies in the area of manufacturing involve steps like conceptualization, product design and development, process planning, raw material management, production planning and control, and quality control.

1.1.3.1. Product Concept (Idea Generation)

The term 'idea generation' is actually a misnomer, since in many companies ideas are rather managed than to be generated. Thus, organizations need to determine sources of ideas and evolve distinct mechanisms to activate those sources. The organizations should endeavor to call upon different ideas. The aim of this stage in the process is to develop a bank of ideas from within or outside the organizations that satisfy the new product strategy and new product development attributes.

Creativity refers to the evolution of new ideas or concepts which may be new to a particular individual, not necessarily new to someone else. It is creativity that leads to the discovery of alternative contemporary product architecture, production systems, or processes which facilitates realization of basic function at the optimal cost levels. Critical evaluation of function by deploying creativity forms the basis of value engineering. It is observed that individuals possess much more skills than they actually perceive. Training and practice can increase the growth of innate creativity. The creative approach banks on the idea generation ability of the problem solver and their ability to embark on the best out of a number of possible solutions.

Innovation means creating state-of-the-art values by providing solutions that satisfy the latest specifications, requirements, diverse customer requirements, and dynamic market demands in a variety of ways. This can be realized by evolving high-capability product designs, production systems, technologies, production planning schedules, services, or concepts that can be pushed rapidly into the markets for meeting societal requirements. Innovation and invention may sound similar, but the two are different from one another. Whereas innovation calls upon deployment of a superior or distinctive idea or concept, invention stands for evolving completely new idea or concept itself.

An invention is a unique or novel practice for realizing drastic changes in existing systems or technologies, which may lead to improvement upon a machine or product, or the development of a new process for obtaining an object or service. A radical invention provides distinct advantages to organizational functions and might not relate to other skills in the same field, thereby completely transforming the particular function or activity.

The author suggested the mechanism for fostering innovation by proclaiming that 'the initial idea may change'. He further added that invention may become simpler, more practical, may expand, or it may even transform into something totally distinct. He concluded that the pursuit of a particular invention can stimulate more inventions.

1.1.3.2. Product Design and Development

At the product design and development stage, engineers conceive and assess new ideas for the realization of tangible inventions, products, processes, or technologies. Product design engineers combine art, science, and technology to create distinct products that benefit society at large. Their creativity is facilitated by digital tools that permit engineers to communicate, visualize, analyze, and actually produce tangible ideas in a unique way, requiring less time and resources than in the past.

Different product design practices focus on different aspects. Some have outlined 'Seven Universal Stages of Creative Problem Solving' for assisting designers to formulate their product from ideas, which include: accept the situation (problem identification), analysis (investigation of problem), define (setting key issues, objectives, restraints), ideate (brainstorming ideas to find solutions for options), select (deciding on selected few ideas), implement (configuring plans, building prototypes), and evaluate (testing product and making improvements).

The adoption of ergonomic principles at workplace have competency to produce bottom-line benefits in mitigating fatigue and injuries at the workplace, besides also providing significant benefits like improving comfort, productivity, health, physical wellbeing, motivation; thereby addressing frequent employee changeovers, absenteeism, and reducing health care expenditures. Ergonomics offers a mechanism for appropriately adopting the work environment to suit the physical requirements and limitations of the personnel, rather than fitting the employees to jobs that can adversely affect the workplace performance and cause personnel discomfort and fatigue. Though ergonomics focuses mainly upon the physical domain, the cognitive and organizational ergonomics domains have been gradually gaining importance due to enhanced automation attributes of modern workplaces. Organizational ergonomics is a new concept that addresses the physical and cognitive domains.

1.1.3.3. Process Planning

Process planning is concerned with assessing the production sequence for obtaining the products using distinct manufacturing specifications, taking into consideration both technological and economic aspects. Process planning requires the skills, experience, and domain knowledge about materials, manufacturing processes, tooling, and associated technological know-how of engineers, depending upon the product design requirements. Automated process planning emphasizes on evolving a single plan that should be optimal in respect of the predetermined product design and development attributes.

Although computer-aided production planning (CAPP) is sometimes used synonymously with production and inventory, it should be used in different context on different parts of the production process. CAPP overcomes several shortcomings of manual process planning like limitations of process planning personnel, inconsistent planning, due to knowledge constraints of the process planner (leading to high time for evolution of process plans), low flexibility, and high cost. Figure 1.4 shows the process plan flow chart.

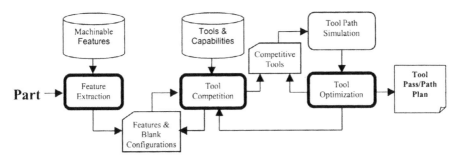

FIGURE 1.4
Process Plan Flow Chart.

Since the process planning demand design information transformation into process attributes and generation of efficient and effective instructions for producing components accurately and precisely, CAPP deploys several computer-aided tools to expedite and precisely control process planning and realize stringent control over transformation of manufacturing resources. CAPP systems employ the knowledge of several competent engineers and product or process designers into an integrated computer-aided manufacturing (CAM) system and computer-aided design (CAD) system that appropriately delivers the detailed process plans and route sheets. Thus CAPP system effective provide a bridge between design and manufacturing and enhances the process rationalization, standardization, besides increasing productivity of process planners and reducing lead time for process planning. A CAPP framework is shown in Figure 1.5.

Material selection is a critical step for designing and obtaining any product, since material properties significantly affects the manufacturing processes, process parameters, tooling, and product characteristics and properties, thereby controlling the cost of the production. The optimal selection of a material for a particular application should be based on considerations of material properties and cost. The material selection is usually dependent upon product design and applications, but sometimes the reverse of it is also true, since the advent of cutting-edge materials open up new applications of the products, thereby revolutionizing product designs.

1.1.3.4. Raw Material and Inventory Management

The production of components is governed by procurement of raw and consumable materials. Material prices are strongly governed by the global and national demand and availability but also by the policies of producer nations of such raw material. Raw material is the essential part of any of

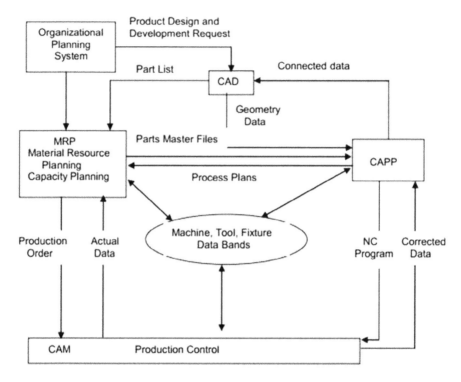

FIGURE 1.5
Framework of Computer-Aided Process Planning.

the manufacturing firms. An organization cannot operate without the presence of the raw material.

Inventory management is a technique for holding and maintaining the placement of stocked goods. This exercise should be taken up at different locations in a production system or at several locations of supply network, to ensure continuous production of goods and commodities, and protect production against any random disturbances of shortage of raw material or consumables. Inventory management is also closely associated with the replenishment lead time of goods, inventory forecasting, inventory valuation, asset management, inventory visibility, the available physical space for inventory, future inventory price forecasting, quality management, returns and defective goods, and demand forecasting.

1.1.3.5. Production Planning and Control (PPC)

PPC is a significant attribute of any manufacturing organization. Manufacturing organizations should develop competencies to forecast, plan, coordinate, supervise, and control the production process. Production planning comprises of evolving key manufacturing attributes like scheduling, routing, dispatching, coordination, inspection, managing machines, operation cycle time, and inventory control. PPC is aimed at streamlining supply and movement of materials and labor, machine utilization and related activities, and for ensuring realization of on time deliveries of quality production. CAM employs integrated computer software for efficiently managing machine performance for ensuring quality and consistent production.

CAM deploys computerized assistance in the smoother and faster conduct of various manufacturing operations, like planning, management, operations, transportation and storage. CAM effectively manages components and tooling with more precise dimensions and material consistency, by processing appropriate raw materials through optimal energy utilization. The green manufacturing revolution can be utilized to the fullest extent in an organization, and if any industry leads with the green manufacturing it can reduce environmental depreciation effects. Several U.S. industries have adopted green manufacturing initiatives by deploying clean technologies and renewable energy, with the U.S. government executing developmental and growth policies.

Robotics involves the design, manufacture, and application of robots; with the computer interface for control, sensory feedback, and information processing. Robotics has gained significant industrial application in situations involving repetitive work, working in awkward positions, handling materials, and sorting applications and above all hazardous situations where human safety at work cannot be compromised at any cost. The contemporary robots have their genesis in bio-inspired robotics and are strongly influenced by human body movement. The science of robotics has grown tremendously in the recent times and has assumed a significant role in industrial; warehouse; and military applications, like bomb defusal, location of IED explosive mines, etc.

Hydraulics deploys pressurized fluid for production, control, and regulation of power. Hydraulic control systems are extensively used in manufacturing industries and provide superior power transmission capabilities compared to pneumatic systems. Both these systems find extensive industrial automation applications in the field of automobiles, agriculture, aviation, construction, defense, fabrication, machine tools, material handling, mining, pharmaceutical, shipping, and the textile industry.

1.1.3.6. Quality Control (QC)

QC measures are aimed at thorough investigation of quality attributes of all factors of manufacturing system. QC approach primarily focuses on key attributes like inspection, process control, and maintaining a complete and up-to-date record of performance of system attributes. The inspection element of quality control emphasizes on conducting measurements, tests, or a visual inspection of certain process attributes in comparison to some predefined standard to evaluate the degree of compliance with specified process requirements. Further, various statistical tools are deployed to ascertain that degree of compliance of the manufacturing process to the specified process specifications and evaluation of process capability, which invites the application of corrective actions for maintaining the process under control. Quality assurance ensures that the quality attributes of production system are maintained and retained over entire manufacturing cycle of production.

Life cycle assessment (LCA) concept emphasizes on critical investigation of entire life cycle of the product/process/service from basic ingredients, processing, distribution, and utilization to ultimate disposal. LCA focuses on mechanisms to critically evaluate environmental impacts associated with all the stages of a product's life from beginning to disposal.

1.2. Competency Issues

Competence is a quality of both people and organizations (Nicolin, 1983). Technological competencies are multi-field highly stable, and differentiated (Patel and Pavitt, 1997). Competencies are broadly classified at the organizational level and employee level (Cardy and Selvarajan, 2006). More experienced and trained individuals will perform better than lesser ones. Also, learning increases reliability and produces fewer surprises (Levinthal and March, 1993). Manufacturing competency is a situation for improved competitiveness. The manufacturing function is a formidable weapon for organizations for achieving competitive superiority (Dangayach and Deshmukh, 2000).

The constituents of competencies (that is, technological, marketing, and integrative competencies) influence company performance (Wang et al., 2004). Manufacturing competencies have an important role in the company's strategic success in order to enable companies to improve their competitiveness. Competitive strategy affects quality and firm performance (Amoako-Gyampah and Acquaah, 2008). Quality and business excellence awards are driving forces in improving the competency of Indian firms in the global market (Dutta, 2007). Small and medium enterprises are considered to be an engine for economic growth all over the world. (Singh et al., 2008).

There are many assumptions for competency models and counters with an investigation of the benefits of leadership competency models for organizations and individuals. A competency model is a more comprehensive model of effectiveness.

In today's competitive world, it is hard for firms to retain sources to respond to new services and new product offerings, such as reducing development time, even if there is a rise in development cost. As competition intensifies, it is important that firms reinforce and protect their sources of competitive advantage and try to compensate for their organizational weaknesses.

1.2.1. Competency Development

Drejer (2001) formulated a framework for competency development in accordance with management in firms. The model was followed by some considerations about practicing competency development in management. Understanding a model as a stable entity and using it as a basis for dynamic development is the key area to focus on.

In present times, companies are facing rapid evolution in integrating new technologies, competitive environments, respect for new environmental and safety regulations, etc. Companies must produce distinctive competencies so as to maintain a competitive advantage (Belkadi et al., 2007). Technology acquisition and integration must be strengthened so as to improve the capability of suppliers in order to receive integrated electronic and mechanical components. To be more competitive, organizations need to develop competencies in a strategic way and thus, develop a strategic architecture for the same. For competence development, a framework, along with competence model, has been presented to form and improve strategies.

Traditional methods are time consuming and expensive. Competency development in sustainable global manufacturing creates experience by showing how specific competencies are identified and how a scenario has been developed. Competencies, in a weak manner, capture and portray different emotions and social interactions in everyday working life. Competency frameworks have an innovative insight into alternative dimensions to improve a firm's performance (Stokes and Oiry, 2012).

In contemporary environments, organizations measure their performance against a set of predefined performance measures. These measures are governed by the ability of the organization for maintaining necessary sets of 'competencies' that assist in the successful execution of its projects. Competencies, being multidimensional and subjective in nature, are difficult to define and measure. A new modeling approach considering prioritized fuzzy aggregation, factor analysis, and fuzzy neural networks has been applied for identifying the relationship between project competencies and key performance indicators. The diverse project competencies are analyzed using factor analysis. A standardized framework and methodology for evaluating the impact of project competencies on project key performance indicators has been established (Omar and Fayek, 2016).

1.2.2. Competency Management

Competency management concerns competency distribution and development. The architecture of the design activities aims to assist competency management through the qualitative feature of the work. For improving business performance, reliable and valid techniques are being developed for measuring and managing competency. Apart from development, advancement of these measurement techniques for end user computing capability is the most pressing need. Knowledge, the major aspect of manpower and assets, is even contained in the simplest tool. In the current market, price is governed more by quality than supply and demand. Production factors and industrial processes are combined to minimize the end product cost, and thus managing competency and improving the firm's performance (Nicolin, 1983).

Competency management, combined with strategic management, represents and evaluates core competencies in relation to processes and products. Competency management (along with the 'value creating network concept' and the diffusion of matrix-based tools) helps in managing the interdependencies between the project domains: organization, process, and product. Competency management experts have underlined the responsibility of design organizations for developing of products in order to satisfy customer requirements (Bonjour and Micaelli, 2010).

During the past few years, important changes have occurred in the automotive industry. With competency management, a new framework has been developed for analyzing the decision of the automakers, whether to develop a new component in-house or outsource it. Competency management led to pass the decision about supplier product to independent, then complementary, of the assembler product.

Hong and Ståhle (2005) identified various key concepts on competency management, which has led to an integrated and systemic view where creating a self-generative and self-renewable organization is a big challenge.

Competency management and development is regarded as a vital tool to improve competitiveness of organizations. Competency management helps to identify opportunities and expand organizations in a global market (Lee, 2006). Indian organizations follow various philosophies (like 5S, TPM, TQM, etc.), however they have not been able to make any substantial improvement in this regard. Competency management and development contributes to a larger extent for the successful implementation of these philosophies. TQM index, competency index, and the 5S index have been analyzed before and after competency-based training in organizations. The manufacturing industry can improve its competitiveness by focusing on competency management and competency-based training along with training on these philosophies (Khanna and Gupta, 2014). HRD interventions have an impact on building of employee competencies, thus improving organizational effectiveness. The HRD interventions undertaken in this study are training, performance management, and career management. A model has been developed by combining various factors. The validity of the model is tested by applying structural equation modeling (SEM) approach (Potnuru and Sahoo, 2016).

1.2.3. Resource-Based Perspective

A resource-based perspective is the competitive advantage that is derived from the valuable, sustainable, and rare resources and capabilities for a firm. A resource-based perspective uncovers the long-term effects of out-sourcing. It focuses on organizational resources for integrating the dispersed knowledge required in developing complex products.

By resource-based analysis, it is argued that cooperative competencies are either socially complex and ambiguous or not based on historical conditions. Cooperative competencies are competencies relevant to knowledge transfer, communication, information processing, inter- and intra-unit coordination, negotiation, and the ability to develop trusting relationships. They can complement technological competencies in dynamic and uncertain industries. Therefore, they can provide organizations with competitive advantage, which cannot be easily imitated.

Cooperative competency and a process-oriented view facilitates the effects of transfer mechanisms, including the adaptation and replication of knowledge transfer performance. Cooperative competencies relate to transfer mechanisms in partnering firms, thus knowledge transfer performance improves. A conceptual model shows interrelationships between cooperative competency, knowledge transfer performance, and transfer mechanisms (Chen et al., 2014). The main focus of companies should be on innovations according to market need due to competition, which can be a tough challenge.

A resource-based view, including core competencies, assist organizations in strategic management. Core competencies and the strategies are based on the ideas that are rare and intangible. There is lack of understanding regarding core competencies and their potential value for the company (Bhamra et al., 2010). A higher priority in manufacturing companies is given to delivery and cost versus flexibility. For accomplishing these objectives, a resource-based view emphasis on shop-floor activities and favors adopting options like periodic reviews, worker training etc. Firms are reluctant in adopting approaches which require either major organizational restructuring or substantial investments.

Individual competencies, composed within self-managed teams, translate into more effective multi-team systems (MTS). A wide range of self-management competencies by team members combine to influence the productivity of a team network (Millikin et al., 2010). The integration of marketing, research and development, and engineering functions (along with strategic thinking and new product development time) is vital for developing competitive advantage. The agility of organizations is a pre-requisite for export involvement (Lim et al., 2006). A resource-based view combines with manufacturing excellence to demonstrate best industrial practices and thus achieving competitive advantage. Frameworks have been developed for the assessment and implementation of manufacturing excellence (Sharma and Kodali, 2008).

1.2.4. Economic Effects

The economic, social, and environmental pressures contribute towards the phases of the automotive life cycle within a global industry. Managing decentralized and dispersed knowledge networks is the key challenge confronting multinational enterprises (MNEs) today. The design of archi-tectures and economics in MNEs for harnessing individual creativity is a critical component of competitive advantage. A competency typology is a methodology for identifying competencies to aid the transition to compe-tency-based logic from task-based HR management. Integrating and imple-menting economics and the competency framework offers guidelines for competitive advantage and managerial insights.

Remanufacturing is a process in which used products are restored to useful form. Reverse logistics is a systematic process for planning, control-ling, and implementing distribution and manufacturing activities (such as the in-process inventory); the packaging of finished goods; and the back-ward flow of raw materials to a point of proper disposal or recovery. Remanufacturing and reverse logistics affect a company's economic and strategic decision making (Subramoniam et al., 2009). Competency frame-work provides the organization with better information regarding selection decisions and understanding of managerial success. Performance ratings, rather than competencies, are influenced by manager's ability to under-stand the business (Sutton and Watson, 2013).

Due to globalization, competitive pressures and the strategic responses of organizations have intensified. Financial performances and marketing outcomes are also varied by sectors. The economic realignment focuses on the internal competitive developments, strategic response to these devel-opments, and the financial and marketing outcomes.

1.2.5. Technological Competency

Technology competencies interact with competitive environments to affect firm innovation and competitiveness. Moreover, different types of techno-logical competencies are required during different types of competitive environment for enhancing a firm's innovativeness. A technological com-petency is described as the ability to use technology effectively, through extensive learning and experimentation, development, and employment in production. Technological competencies have emerged strongly during the twentieth century in their corresponding fields.

Geels (2001) described technological competencies as long-term and major technical changes in the way of satisfying customers' needs. Technological insights have three levels: socio-technical landscape, socio-technical regime, and technological deliberation.

Technological competency and research and development focus on the creation of competencies and new knowledge and the benefits that firms can reap from collaboration. For acquisition of competencies, a perfect balance between the firm's technological development and the collaboration's strategic orientation is required (Mothe and Quelin, 2000). Information technology competency (ITC) plays a critical role in learning various competencies. In fact, learning competencies mediate the relation between ITC and commercial success of innovation (CSI) as CSI is directly related to competencies (Fernández-Mesa et al., 2014).

The technological competencies are, to some extent, multi-field rather they become more with time and competencies beyond their product range. These competencies are highly differentiated and stable, with both the direction of a localized search and technology profile strongly influenced by organization's principal products (Patel and Pavitt, 1997). Technological competencies have the ability to improve products, and thus the customer base, as they create and use technologies effectively. Competency development plays an important role in enhancing industrial competitiveness. Risk, status, uncertainty, observability, disruptiveness, pervasiveness, and centrality are technical characteristics that influence the learning modes selected by an organization. Technological competencies have an impact on organization performance and possess all the prerequisite competencies required for a given technology-product-market paradigm as the organization enters or remains over time in market. Consequently, high-tech entrepreneurial firms must learn, acquire and develop competencies in response to the changing requirements of industry products (Linton and Walsh, 2013).

1.2.6. Competition

Success and survival in today's tough environment depends on competitiveness. Competitiveness is achieved through the integration of efforts between various functions of organizations and the use of advanced manufacturing technologies (AMTs). AMTs have a vital role in flexibility and quality improvements in manufacturing organizations. Technology is also a competitive weapon as new products and processes with complexity and higher accuracy are introduced and changes in customer needs and expectations. AMT helps manufacturers in enhancing financial and quality performances and thus, improving competitiveness (Dangayach et al., 2006).

Survival in today's highly fragmented market requires firms to focus on factors such as adherence to standards, quality, rapid response, and innovation as the basis for competitive advantage. In order to meet these demands, firms are deploying innovations, including the reconfiguration of their internal organization, advanced equipment, and their external relationships. Competencies in innovations is the most pressing need for organizations, and most importantly, balancing the existing competencies to survive in competitive world. Organizations need to master the competencies in innovation so as to exploit the various phases of an innovation process (Janssen et al., 2014).

The manufacturing industry plays a key role in a country's economy. Also, it adds high value to exports, generates jobs, and contributes to political and social stability to achieve higher profits. Functional competencies are important to an organization's performance, as they help the companies in improving their competitiveness.

The strategic automobile shredder residue (ASR) recycling model helps vehicle recyclers in improving their economy, as there will be a decrease in disposed ASR quantity with an increase in recycling. Due to such factors – such as increases in oil prices, advances in technology, and government regulation of emissions – the automobile market has entered into a period of uncertainty. The existing manufacturing industries try to develop and patent alternative fuel vehicle (AFV) technologies (Wee et al., 2012).

Competencies are personnel attributes that offer significant organizational benefits in realizing significant performance enhancement. Competency means clearly establishing job requirements and should not be used for making a distinction between competent and not competent personnel for the purpose of layoffs. Developing competencies means to inculcate skills that facilitate personnel to realize significantly better outcomes from the workplace, as it is job effectiveness that really matters for determining the success of an organization. Although conventional approaches to competitiveness tend to be of limited scope, some methods and models would take the more comprehensive approach. The corporate core competence concept advocates for developing an appropriate understanding of internal and distinct factors for depicting an organization's differentiation and specialization attributes. The literature reveals that inculcating competencies in organizations can facilitate them to accrue strategic success in the fiercely competitive global marketplace.

2

Strategy and Its Aspects

2.1. Strategy

Strategy finds its genesis from the Greek word meaning 'generalship' and aims at inculcating leadership attributes for various initiatives in organizations. Strategies are usually aligned with overall organizational objects and should holistically take in consideration available organizational resources and constraints. Top management frequently deploys various strategies to accomplish its organization's goals and objectives. While objectives underline the aims of an initiative measured by success measure, strategy depicts the path to success. Therefore, strategy provides the course of action for the realization of such goals and objectives (Flint, 2000).

Strategy means deliberately selecting different sets of activities for creating a unique mix of values (Porter, 1996). Corporate strategy provides an overall plan for the firm, since production is separate from the purchase and consumption of these goods (Jones and Parker, 2004). Major emphasis in various models, theoretical concepts, and frameworks has been given to manufacturing strategy (MS). There seems to be a relation between production competency and business competitiveness, thus leading to improved strategies.

Strategic success embodies formulation and execution of key organizational objectives, taking available resources into account. Strategic thinking involves effective creation and execution of distinct business understanding and opportunities aimed at realizing sustainable competitive advantage. Strategic planning initiatives can be carried out by individual alone or cooperatively for providing favorable organizational benefits.

2.1.1. Strategic Agility (SA)

SA describes the aptitude of organizations to attain business competitiveness by strategically evolving and imbibing innovativeness into design, manufacturing processes, services, and related business functions. Agility has been envisioned as an indispensable element in an organization's ability to imbibe change quickly for long-term pursuit of growth and competitiveness. SA is a multidimensional concept that focuses on organizational capabilities to create new market opportunities, establishing effective mechanisms for evolving business process transformations rapidly to produce innovative products and services for realizing business-related competencies.

SA encompasses various forms of agility, namely operational, customer, partnering, portfolio, business process, supply chain agility, etc. Thus, SA demonstrates the organization's keenness to both anticipated and sudden changes, by optimally deploying resources and knowledge to evolve innovative solutions for meeting short-term and long-term competitive edge. Strategic sensitivity, collective commitment, and resource fluidity have been outlined as critical dimensions of SA.

SA and learning has three dimensions: process adaptation and experimentation, collaborative technology sourcing, and proactive technology posture. These aspects help organizations in gaining competitive advantage (Ahmad and Schroeder, 2011). To be more competitive, organizations need to operate in a strategy-driven way and develop a strategic architecture to enable them for developing necessary core competencies. MSs contributes to company level competitiveness. Intuitively, it seems obvious that a smoothly running manufacturing system will have significant influence on firm performance.

Alsudiri et al., (2013) described the factors for SA, project management, and business strategy. A framework provides a relation between business strategy and project management. This helped the companies in implementing business strategies by embedding their projects in the overall strategy.

SA and development are considered the engine for economic growth and improving competitiveness of SMEs (small and medium enterprises) in the global market (Singh et al., 2008). SA gives a new dimension to thinking about strategic entry barriers whereby competitor's decisions are altered by government rules. Companies attack competitors by modeling government rules so as to create a misalignment of competitor's transactions and governance structures. A firm's governance structure will be established by the guidance from the attributes of the transaction (Averyt and Ramagopal, 1999).

Regional strategies are alternative, potentially superior solutions for globally integrated and locally responsive approaches. Companies follow a diverse path for regionalization and globalization. Such an approach involves a cluster of nations that reveal similar market conditions and homogenized regional consumer needs, while minimizing adaptation costs. Regional strategies are associated in evolving a firm's global strategy (Schlie and Yip, 2000).

Financial performance is often used to measure firm performance. There is need for an integrated framework to assess manufacturing flexibility and firm performance. Moreover, SA and thinking is reckoned to be an integral parameter towards manufacturing flexibility and competitiveness (Mishra et al., 2014).

Corporate social responsibility (CSR) is more important to managers, still CSR is considered as an informal structure of corporations. According to economic conditions, CSR has an impact on corporate reputation and customer satisfaction in the automotive industry. Although, there is no correlation between corporate reputation, economic responsibility, and customer satisfaction. Thus, to increase efficiency in production, the industries need to be strategically agile, so that managers are open to new perspective with most basic and important responsibilities (Hanzaee and Sadeghian, 2014).

A CSR competency framework supply CSR skills and competencies for managers and practitioners. Large companies use these attributes with regard to attitudes, knowledge, and practical skills. The organizational competency framework of strategy for integrating CSR and its associated skills into mainstream business has been provided (Shinnaranantana et al., 2013).

2.1.2. Strategy and Business Performance

Strategic thinking has a significant impact on strategy formulation and strategic actions thus affecting business performance. Strategic thinking models consist of conceptual ability, visionary thinking, analytical ability, synthesizing ability, objectivity, creativity, and learning ability. This set of abilities and skills are termed 'strategic thinking competency.' The strategic thinking competency model offers a framework for developing strategic thinking which contributes to better strategy and better business performance. This model is applied for designing training programs to develop better strategic thinkers (Nuntamanop et al., 2013).

The major challenges of upgrading technology, reducing costs, and building product quality are common to SMEs globally. Although Indian SMEs lack the capacity in product design and development capabilities, government policies have a prominent impact on development of strategy for competitive nature of SMEs globally (Singh et al., 2010). A framework shows the relationship between the process of manufacturing and the strategic thinking. Various forms of manufacturing strategy processes and their impact on business performance have been provided. Strategic thinking competencies facilitate process innovation, as organizations can enhance their performance by their innovative efforts for developing and strengthening relevant strategies (Tarafdar and Gordon, 2009).

Through the use of mid-segment cars, the industry's concept, evolution, physical aspects, and contribution to the growing need of the segment have been highlighted.

Outsourcing along with strategic thinking influences sustainable growth in their organizations. At the technological level; machining, and composite technologies and at the product group level; structural parts and brake system parts are core competence factors for an organization. Various criteria – such as operation effectiveness, safety, technological feature, cost effectiveness, usage quantity, procuring sufficiency, and workforce – are effected by core competencies, thus core competencies in an organization effects strategic outsourcing (Demirtas, 2013).

2.1.3. Role of Management

Management is envisioned as a significant contributor of production besides money, machines, and materials. Peter Drucker described innovation and marketing as key attributes of management initiatives. Management comprises of accomplishing closely interrelated functions of evolving planning, controlling, organizing, corporate policy, and directing a firm's resources for realizing strategic goals. Management control is defined as:

> Management control stands for a coordinated approach by an organization to evaluate business performance against to predetermined standards, objectives to ascertain their stringent compliance and ensure optimal utilization of personnel and other corporate resources for realization of organizational goals.

Performance improvement initiatives call for evaluating and comparing the actual system performance against desired performance measures. A situation depicting reasonable gap between actual and desired performance levels provides the organization with an opportunity for taking up effective performance improvement initiatives. Performance management and improvement process should be based on three attributes of effective *performance planning, performance coaching*, and *performance appraisal.*

Management coordinates the efforts of people to accomplish various objectives by utilizing available resources effectively and efficiently. A competitive advantage is a management concept that describes attributes which allow an organization to outperform its competitors. There may be different competitive advantages including customer support, a distribution network, a company's product offerings, and cost structure. For a competitive advantage, customers should be able to distinguish differences between one company's products and services and their competitors. This difference exists because of a capability gap between a company and its competitors. The competitive advantage leads to technological innovation and thus, leading to better strategy formulation and management (Coyne, 1986).

Technological innovation at a products or processes level affects the environmental and economic condition of the industry. For achieving sustainability, a change in processes must be undertaken at the system and functional level. A product service system (PSS) is designed for competitiveness with lower environmental impact and satisfying customer demands. PSS involves developing, maintaining, obtaining, and improving products and services in response to market opportunities. PSS is assessed against expanded set of key evaluative criteria: changes in institutional and infrastructural practices; changes in vehicle ownership structure; evidence of 'higher-order' learning amongst stakeholders; changes in vehicle design, manufacture, and end-of-life management; and changes in modes of producer and user interactions. Information and communication technologies (ICTs) are used by most organizations to offer more sophisticated innovation options (Williams, 2006).

Firm competencies, such as, new product development (NPD), marketing management, and business networks have an effect on strategic management of an organization. Business networks are a kind of competency that covers inter organizational networks, government relationships and research and development partnerships. NPD consists of product process innovation research and development capabilities, while marketing management includes branding, information management, promotion, and distribution channels (Wang and Lestari, 2013). Loan amounts play an important role on the forecasting of automobile sales. There exists a relation between automobile production and automobile loan amounts, as automobile production through sales varies according to different loan conditions and amounts. Thus, loan amounts from banks have an impact on the strategic management of an organization (Yildirim et al., 2012).

Preventive maintenance is carried out on equipment to avoid its breakdown or malfunction. It is a routine and regular action taken on equipment. Preventive maintenance can be extended effectively only if maintenance engineering instruments are adopted to assure that preventive maintenance programs harmonize well with production schedules and with the actual state of manufacturing equipment. The maintenance performance of an organization leads to growth in firm's performance. Thus, it can be said that preventive management affects an organization's competencies and strategies in the present tough environment.

2.1.3.1. Strategic Management

Strategic management is the formulation and implementation of the initiatives and major goals taken by company's top management, based on consideration of resources and environments in which the organization competes. In today's world, the main feature of the economy has changed from an individual's effort to teamwork. In an inter organizational network, competencies are built by each firm and are influenced by strategy formulation (Fleury and Fleury, 2003). Intense competitiveness and rapid developments in manufacturing technology and information technology have led to turbulent and uncertain marketplaces throughout the world. Advances in biological and physical sciences and the shortening of product life cycles to some extent lead to uncertainty. There has been rapid transition in production systems to new organizational forms due to radical changes in technologies, competition, and marketplaces. The manufacturing function and strategic management is a formidable weapon to achieve competitive superiority. A model links the action plan of firms and the manufacturing competitive priorities (Dangayach and Deshmukh, 2000).

A company's internal and external variables are linked regarding strategic management, which leads to improved marketing performance. Strategic management includes human resource management, marketing strategy, and strategic thinking. Marketing competency substitutes strategic management, but a formalized organizational structure hampers it. Along with this, the attitude of organization's management towards risk taking and the emphasis of CEOs on strategic thinking also slows down, due to the formal structure of the organization. Strategic thinking is positively related to centralization in the organizational structure. Market performance improves with enhancements strategic management and technological turbulence (Moom, 2013).

Strategic management elements in the operations unit and business strategies are corelated. A model has been presented that associates organization's performance with the challenges in the strategic management field (Shavarini et al., 2013). Managerial cognition along with corporate and individual values have an impact on strategic management. Strategic management and competencies are important to a firm's performance. Along with management, psychological elements provide an understanding of the strategic thinking and decision-making process (Steptoe-Warren et al., 2011).

Strategic management along with quality management affects the quality as a source of consistent competitive advantage. Quality competency plays an important role in sustaining advantages in today's highly competitive world (Zhi-Yu et al., 2006).

2.1.4. Knowledge Transfer Management

Knowledge, technological skills, and capabilities constitute a major part of a firm's competitiveness. For the creation of knowledge that will be transformed into new products and services, firms have started to form knowledge-based strategic alliances, thus a new form of competition has been created. Nevertheless, this creation and transfer of knowledge through inter-firm cooperation has proven to be quite difficult. The issue of knowledge transfer management is more important for firms lacking in technological capabilities. Strategic variables related to knowledge and flexibility affect an organization's performance. A framework has been presented for capturing knowledge and information from external sources and the process that organizations use to assimilate transform and use this knowledge (Fernández-Pérez et al., 2012).

Knowledge management helps in dealing with the complexities of social behavior in organizations. Knowledge plays an important role in firm's performance, as strategic actions are planned according to knowledge and knowledge management, which involves workers as well as managers (Moore, 2011). Along with knowledge management, habitual expressions also form a part of the production process. Individuals use various habitual expressions for conveying their desires for product forms. Even oral styling procedure is described to a designer, who progressively sketches the product as it is described (Chang et al., 2005). Multi-team systems are greatly affected by individual capabilities in self-managed teams. Self-management competencies influence the productivity of a team network. Multi-team systems are comprised of teams which widely practice self-management and knowledge management strategies to attain higher productivity. Multi-team systems with consistent self-managers are the most productive ones (Millikin et al., 2010).

Knowledge transfer and management has led industries switch to re-insourcing. Besides outsourcing, re-insourcing helps organizations with decision making in manufacturing strategy and in achieving various underlying motives. Re-insourcing implementation is gaining momentum, especially during situations of financial crisis (Drauz, 2014). In modular assembly, automobile manufacturers can either outsource the units for assembling or assemble them internally. Module assembly units' performance is affected by ownership and location. Even the internal and customer side conditions of different organizations greatly affect module assembly units' performance (Fredriksson, 2004).

A fit manufacturing framework helps manufacturing companies to operate effectively and become economically sustainable in a global competitive market. A fit manufacturing paradigm (through effective marketing and product innovation strategies) integrates the manufacturing efficiencies, agility, and new markets to achieve long-term economic sustainability (Pham and Thomas, 2012).

2.1.5. Strategy Management and Technology

There is a strong link between technology strategy and business management, but organizations can have different approach for technology acquisition and development. Strategic management and technology is practiced in various organizations in the auto component industries. Effective strategic management and technology, or strategy technology management, contributes to faster technology absorption and improved firm performance (Sahoo et al., 2011).

Certain business adaptations are required while entering a new market for improved strategic technology and management. MATCH is a new conceptual framework that assesses the strategic management and potential value creation in relation to the business model (Sleuwaegen, 2013). Strategy management and technology is affected by the techniques for organizational learning process. Cognitive maps of the Indian automobile industry are used for understanding the organizational learning process. A simulation experiment, in light of organizational learning, was performed on an uncertainty-based cognitive map for generating future scenarios (Srinivasan and Shekhar, 2000).

Maintenance quality function deployment (MQFD) is a feasible model for nourishing the synergic benefits of TPM and quality function deployment (QFD). Successful implementation of MQFD leads to an enhanced firm performance as by this model strategic thinking, which involves strategic management and technology, gets improved. TPM implementation leads to the enhancement of employee competency. It also shows how competency has been improving by implementing TPM. The study uses analytical hierarchical process (AHP) for analysis. This study integrates TPM implementation with employee training, empowerment, teamwork, compensation, and management leadership in a theoretical model for studying employee competency within the framework of a management system (Maran et al., 2016).

Strategic technology management involves concurrent engineering and lean manufacturing towards new product development. Moreover, a high degree of similarities exists between concurrent engineering and lean manufacturing. With lean manufacturing, organizations are able to develop new products faster and better with less cost than conventional companies. The benefits of lean manufacturing principles go much beyond frequent deliveries and inventory reduction (Meybodi, 2013).

Entrepreneur training and education ensures the improvement in the performance of micro firms. An investment in entrepreneur training and education is especially relevant to firms which are highly competitive. This implication leads to improved competencies and strategic management. A positive association between a business training program and company performance exists based on four performance variables: organizational improvements, better job satisfaction within a company, increased exports, and an increased number of employees (Yazdanfar et al., 2014).

Performance management and measurement (PMM) techniques and tools, in recent years, have gained interest, as implementation of PMM yields many advantages. There are number of benefits experienced by organizations after introducing PMM. It leads to better strategic formation, since strategic technology and management improves, thus improving competencies. Management should make PMM advantages explicit before starting their implementation and stress these advantages during and after implementation. With this, the commitment of organizational members' will boost up PMM and its successful use (Waal and Kourtit, 2013).

Firms with strategy technology management use open innovation practices and core competencies, whereas firms with diversification strategies use managerial practices. The framework gives useful indications about the dependency of firm performance on competitive strategy and open innovation. Firms with an innovative strategy invest more on technical skills and core competencies. With the integrated model developed, links open up innovation, innovation performance, and the company's strategy in SMEs (Crema et al., 2014).

2.1.6. Teamwork

Teamwork is an essential asset for any organization since teamwork has the potential to produce incredible results in any organization. Teams may comprise of interdependent group personnel with differentiating skills and knowledge levels from varying fields, that are devoted to realization of a given task objective or overall organizational goals. Each team member or associate might contribute marginally towards the task objectives, but overall team participation leads to the achievement of greater goals, beside the enlightenment of personnel working in the teams. Teamwork assumes a significant contribution in ensuring the successful completion of many projects. For realizing meaningful results, team associates should work closely with clearly defined roles and responsibilities and should be considered equally responsible for their team's success or failure.

2.1.7. Administration

Administration means effectively managing an organization or business to ensure the realization of consistent business results and the long-term competitiveness of the organization. The administration function encompasses decision making to ensure the smooth execution of business operations, besides ensuring efficient organization of human and other related organizational resources, as well as to realize organizational objectives.

Administrators are entrusted with the task of performing common set of functions to meet the organizational goals. Henri Fayol has suggested that the key elements of administration are planning, organizing, staffing, directing, controlling, and budgeting. Administrators should possess bare minimum basic skills, involving management, technical, interpersonal, and conceptual skills.

2.1.8. Interpersonal Relationship

An interpersonal relationship is the coherent working of many individuals to work collectively in a team towards fulfilling an organization's goals.

Employees need to be committed to each other and organizational objectives to create a healthy and congenial environment at work. These relationships may be formed in the context of social, cultural, and other related task influences. Interpersonal relationships create overall organizational transformations and create an environment of mutual trust and motivation within an organization. Teamwork requires that personnel working together should appreciate contributions made by each individual and must share a special bond, so as to be able to perform to the best of their abilities.

3

Manufacturing Competency and Strategic Success

3.1. Manufacturing Competency

The literature highlights that the organizations offering flexible production routines, exhibiting product innovations and timely responsiveness, have reaped tremendous business success through effective coordination and redeployment of internal and external competences (Teece et al., 1997). The organization's short-term competitiveness can be realized through strategic success attributes of different products, while the long-term competitiveness can be attained through the competency of an organization to deliver cost effective innovative products with a high degree of flexibility (Prahalad and Hamel, 1994).

Manufacturing strategy is influenced by competitive strategy, and both affect firm performance in the manufacturing industry. There exists a significant relationship between manufacturing and competitive strategies of quality, flexibility, cost, and delivery. Firm performance, to a great extent, is affected by quality. An emphasis on quality provides a means by which organizations can mitigate the effects of a competitive manufacturing environment (Amoako-Gyampah et al., 2008).

Manufacturing competency, along with competence management, is comprised of management, leveraging, building, and deployment of operational competencies and strategies. The causal relationship between them depends on the way competencies are embedded in individual and organizational resources. Competence management and agile management are interdependent to some extent. Both global and strategic manufacturing competencies play an important role in an organizations' performance and technology development. Manufacturing competencies support strategies to enhance the competitiveness of an organization in global market (Chaiprasit and Swierczek, 2011).

Three concepts of manufacturing competency (that is, organizational capability, administrative heritage, and core competency) are interrelated with economic condition and strategy formulation. A resource-based view complements economic analysis, and both are vital for complete understanding of global strategy. A manufacturing competencies framework consists of competencies, such as research and development, production, information systems, HR, and marketing competencies. There is a considerable impact of functional competencies on organization performance. Also, marketing competencies, research and development competencies, and production competencies are key determinants of customer satisfaction and efficiency (Masoud, 2013).

Manufacturing competency and competence-based management have led to major changes in the field of strategic management. As there is an link between a design management field and a value creating network, changes in one lead to changes in another. Therefore, design managers need to use these fields together in order to improve the company's competitiveness. Design organizations contribute strongly to the organization's core competence (Bonjour and Micaelli, 2010). In-house development and external acquisition, along with organizations' strategic orientations, are interrelated. Low- and high-performing firms differ regarding their strategies and methods for competency development and acquisition. The difference is related to the combination of various integrated and discrete factors.

Technology adoption affects operational competitiveness in international manufacturing organizations. The mechanism developed provides a deep understanding of trust building for improving operational competitiveness. The managers should focus on their actions, which in turn will help to improve their firm's competitiveness (Kristianto et al., 2012).

Knowledge and manufacturing competency variables affect international performance. Small firms expand into international market by examining the effects of absorptive capacity on the relationship between organizations' networks and knowledge competencies. Drivers of knowledge competency help in performance and early expansion of the firm in a global market (Park and Rhee, 2012). Drastic changes in customer expectations, technology, and competition have created an uncertain environment. In order to control this situation, manufacturers are seeking to enhance competencies (especially manufacturing competency and flexibility) across the value chain. Manufacturing flexibility is the ability of an organization to produce a variety of products according to customer demand and expectations while maintaining high performance. Moreover, it is a critical dimension of value chain flexibility. It is strategically important for enhancing competitiveness and winning customer trust. The existing literature describes a framework for exploring the relationship among flexible capability, flexible competencies, and customer satisfaction (Zhang et al., 2003).

3.1.1. Competency-Based Business Performance

Technical activities are vital factors in the globalization of multinational firms. The intangible nature of technological assets suggests that

R, D and E (Research, Development and Engineering) activities should be managed strategically, sometimes favoring centralization and, at times, decentralization. Strategic management of R, D and E activities leads to improved competencies and hence enhanced business performance

Companies should maintain a higher degree of rationality among order-winning criteria, competitive priorities, and improvement activities due to the customer pressures and competitive market. The competitive priorities of Indian companies are improving process and product quality as well as on-time delivery. Companies try to enhance competitive priorities by implementing manufacturing strategies from, total quality management (TQM), material requirements planning (MRP), just-in-time (JIT), and statistical process control (SPC). Manufacturing efficiency and quality are significant criteria in today's industry towards competency and business performance (Laosirihongthong and Dangayach, 2005). These challenges provide the basis for new and advanced manufacturing strategy which aims to create economically sustainable manufacturing organizations. Along with these challenges and their corresponding technologies, strategies, and competencies are improved, thus achieving better firm performance (Thomas et al., 2012).

Competencies influence the performance of different types of services of industries. Services have often been classified as the predictability of supply and demand, maintenance level, degree of labor intensity, method of delivery, and level of customization. These dimensions propose different types of services as service factories, service stores, service shops, and service complexes. Along with competencies, services often improve by having proper strategic management, thus leading to improved competitiveness and firm performance (Davis, 1999).

Cloud manufacturing is a manufacturing model which is service-oriented, customer-centric, and demand-driven. It presents a strategic vision for the field which is influenced by competencies. Key commercial implementations are critical to the production of cloud manufacturing, including industrial control systems, automation, service composition, flexibility, business models, and architectures. Further improvements in business performance can be made by having higher competition and improvements in the areas of industrial control systems, business models, flexibility enablement, and cloud computing applications in manufacturing.

There is a relation between export performance, competencies, and manufacturing strategy of manufacturing SMEs. By adopting the manufacturing strategy, SMEs would gain a competitive advantage over their rivals and reap higher exports. They could also seek out opportunities and anticipate future threats for further expansion in the global markets (Singh and Mahmood, 2014).

3.2. Strategic Success

Strategy is the direction of the organization. It ideally synchronizes with its resources and changing competencies. It signifies deliberately evolving distinct initiatives to accrue a unique mix of value (Porter, 1996). Strategic success embodies formulation and execution of key organizational objectives, taking available resources into account and by making holistic considerations of internal and external organizational competitive environments.

Strategic planning must yield the necessary data and facilitate the organization to execute management efficiently in realizing business potential. Strategic planning assists executives to take key business decisions in light of current assumptions and understood practices. Strategic thinking involves the effective creation and execution of distinct business understanding and opportunities aimed at realizing a sustainable competitive advantage. Strategic planning initiatives can be carried out individually or cooperatively for providing favorable organizational benefits. Strategic thinking calls for acquiring and evolving strategic innovative ability by examining available organizational resources and confronting common knowledge to make better decisions. Recent strategic thought focuses towards the conclusion that the critical strategic question is not the conventional 'What?', but 'Why?' or 'How?'

Interdependencies between process, organization, and product have been managed by using the diffusion of matrix-based tools. Design management, along with strategic management, represents and evaluates design core competencies in relation to organizational, product, and process architectures. Small business owners should understand strategies required for growth, as they may lead to unintended consequences, like the demise of their firms. Strategic formulation that enables larger organizations to survive and grow do not necessarily have the same effects for smaller ones. Certain boundary conditions exist for the effectiveness of strategies on firm performance in terms of size (Armstrong, 2013).

3.2.1. Competitiveness

Competitiveness is the ability and performance of a firm. Competitiveness may be short-term and long-term. Short-term competitiveness lasts through one business cycle whereas long-term competitiveness lasts over more than one business cycle. The 'unthreatened competitive advantage' is the noticeable advantage which had the potential of lasting over the entire length of the predictable future (Flint, 2000). Manufacturing capabilities affect the competencies and strategy formulation as well as competitiveness. The concept of absorptive capacity that has been provided to demonstrate that manufacturing capabilities are associated with operational performance, and to show that for firms whose main competitive priority is operational performance, the association between manufacturing capabilities and operational performance is the strongest. Manufacturing capabilities have a significant impact on integration of the customers. Manufacturing capabilities have a direct contribution to firm performance (Haartman, 2012).

Maintenance is becoming more challenging in the present dynamic business environment as it leads to competitiveness in market. Maintenance management systems focus on improving productivity, equipment effectiveness, environmental issues, and workplace safety. Maintenance practices in manufacturing companies have good prospects as all practices are at a high level and affect business performance to a larger extent.

Functional strategies of manufacturing, market orientation, and human resource management have an impact on competitive strategy. They affect performance as it differs among family and non-family firms. There exist significant relationships between the functional strategies and competitive strategy. An HR participation strategy, flexibility, and a delivery strategy have a significant impact on profitability for family firms, but not for non-family firms. Cost manufacturing strategy and market orientation is related to cost leadership strategies for both family and non-family firms (Amoako-Gyampah et al., 2007).

Performance measurement systems have an impact on business performance and, by extension, competitiveness. They include the use of integrated measures related to the perspectives of the balanced scorecard and the stage of development. With performance measurement systems, the organizational performance was higher than that of their competitors. Flexible, integrated performance measurement systems for greater administrative intensity and higher level of formalizations leads to an increase in organizational efficiency. Certain quality management practices have a significant effect on customer satisfaction and productivity improvement. Hence, these practices increase competition in the market and lead to improved firm performance.

Almost all countries try to bridge the gap in competitiveness between domestic and global firms. The fundamental characteristic of product development strategies is based on variety, and leads to introduction of the new perspective of product platforms. Furthermore, relating to suppliers, the black box method has improved. Two approaches possible towards product line management are called 'static' and 'dynamic.' The isomorphism between design and production activities enables the lean production system analysis to be applied to design. The evolution of the product development model adopted by organizations has relation with the strategy of product variety. The strategy and organization at the multiproject level is considered the most promising (Muffatto, 1998).

Industrial competitiveness is vital for nations having export-oriented industrialization policies. The competitiveness of an industry, being a complex process, can be analyzed from various perspectives. The aggregate performance of many organizations can reflect their competitiveness. Based on theories from operations and strategic management, an AHP-based model has been presented to explore the varying degrees of significance of the drivers and indicators of industrial competitiveness. The model identifies the degree of importance of organizational performance indicators when assessing industrial competitiveness. Further, it helps to evaluate the factors that are important for companies to perform better.

3.2.2. Company Enactment

The automotive sector is flexible in the development of strategies. These strategies (related to competency development, investment, quality, and cost) are significantly correlated with competitiveness. A growth-supportive environment, the shortage of technical manpower, and raising funds from the market are major constraining factors; whereas delivery time, quality, and cost are the main pressures on the automotive sector. Strategies along with the different dimensions of competitiveness have a relationship with these major factors. Organizations should make investments for developing new competencies and address quality improvement and cost reduction (Singh et al., 2007).

Corporate financial performance (CFP) is affected by corporate social performance (CSP). By increasing productivity and lowering costs, CFP might also be affected by the impact of perceived CSP on customer satisfaction. CSP in terms of environmental performance, labor practices and community development contributes to consumer satisfaction. However, determinants outside the empire of CSP – such as product quality, perception of service, and understanding the brand – were highly important for customers. The overall importance of CSP for customer satisfaction suggests that in the automobile industry, CSP may contribute indirectly by increasing consumer satisfaction, as well as better financial performance, directly by increasing productivity and reducing costs (Reijnders et al., 2012).

According to Kassahun and Molla (2013), the business process reengineering complementary competences (BPRCC) is composed of the business process reengineering complementary transformational competences (BPRCTC) and the business process reengineering managerial competences (BPRCMC). The BPRCC has an impact on company performance, as it leads to improved strategies, thus improved business competencies. The BPRCC with measurement instrument for public sector of a developing economy has been modeled.

The impact of an advanced, even authoritative structure on an organization's execution and development encourages the accomplishment of its worth and, in addition, has an immediate effect of administrative abilities on an organization's execution (Verle et al., 2014).

Agile manufacturing, a manufacturing paradigm, is the winning strategy to be adopted by manufacturers for drastic performance improvements to become leaders in highly competitive market and changing customer needs. 'Agility' basically refers to the efficiency with which an organization accommodates changes in product and process. The drivers of agility and competitive advantages have emerged because of changing manufacturing requirements. The need for achieving competitive advantages of manufacturing and without tradeoffs is essential for an agile paradigm.

With an increase in the turbulent and dynamic nature of competitiveness, there is a tendency to understand organizations in terms of the effective and efficient use of capabilities that creates consistent industrial performance. The main concern of contemporary strategic management is the development of more effective methods for management of knowledge and intangible resources. The relative influences of three constituents of core competencies (that is, technological, marketing, and integrative competencies) have an impact on strategic competency. Market turbulence and technological turbulence are moderately significant, but exist among all the relationships between major constituents of organization performance and core competencies. Market turbulence regulates the relationship between integrative competencies and organization performance (Wang et al., 2004).

Organizational learning capability has an impact on product innovation and the performance of SMEs and this relationship is facilitated by design management capability. Interplay between product innovation, design management capability and organizational learning can be very useful in better understanding for improving innovation performance. Design management, being dynamic in nature, emerges from learning and allows the company to adapt to environmental changes. The industry-based efficiency evaluation provides management with information about the relatively best practice and locates the relatively inefficient sectors by comparing them. Based on the literature, more perspectives – such as service activities and financial power – contribute to the four main competitive priorities of quality, flexibility, cost, and delivery.

3.3. Objectives

Taking into account the literature survey, the need for this work arose because the impact of manufacturing competencies on strategic success of a firm have not been yet addressed. For this purpose, the present book has been designed to investigate and suggest manufacturing competencies that contribute to the strategic success of the automobile manufacturing industry. The present work has emphasized on the following objectives:

1. Synthesizing the concept of strategic success in automobile industry
2. Exploring manufacturing competencies in automobile industry

3. Analyzing the impact of manufacturing competencies on strategic success in automobile industry

Based on the above objectives following issues have been explored in this work:

1. Strategic success in the context of automobile industry has been defined.
2. Manufacturing competency and its components have been elaborated.
3. The various aspects of strategic success to manufacturing competency have been correlated.
4. The impact of manufacturing competency on strategic success has been analyzed and modeled using suitable qualitative and quantitative techniques.

3.3.1. Methodology Adopted

For accomplishing the objectives, the following methodology has been followed:

1. A detailed literature review has been carried out to ascertain the significance of manufacturing competencies and strategic success.
2. A survey of several automobile manufacturing units have been completed through a specially prepared questionnaire for understanding and assessing the current situation.
3. Suitable qualitative and quantitative techniques have been employed to correlate manufacturing competencies and strategic success.
4. To authenticate the findings of the survey, these findings have been followed by case studies in selected automobile manufacturing units of North India.
5. The results of survey and case studies have been synthesized to come out with a suitable model.

3.3.2. Phases of Research

The study has been carried out with the objective of developing effective factors for manufacturing competency and strategic success in the North Indian automobile manufacturing industry. Considering the complexity of theme and taking into account the fact that such studies can be carried out primarily by closely analyzing the approaches adopted by various organizations and results thereof, it was considered appropriate to carry out the study under the flexible systems methodology (FSM) framework.

The three basic components of FSM are actor, situation, and process. The 'situation' is to be managed by an 'actor' through an appropriately evolved management 'process', which recreates the situation. The 'actor' forms a part of the 'process' as well as the 'situation'. The research involves the following phases:

(a) **Clarifying the Context**
A detailed literature review has been conducted regarding competency and strategy factors adopted by manufacturing organizations worldwide, from time to time in the past and issues involved with the same. The evolution of processes has been studied through various stages of manufacturing and strategy practices along with their relevance and shortcomings. The literature review illustrates tools and techniques employed in implementation processes of these factors and the potential benefits accrued by the Western world through effective competency-strategy implementation programs.

(b) **Understanding and Assessing the Situation**
A survey of large number of manufacturing organizations (118 manufacturing units) have been completed through a specially prepared questionnaire for understanding and assessing the present situation regarding the manufacturing competencies of Indian entrepreneurs. The survey design and analysis involves the following steps:

  i. The design of a questionnaire on various aspects of competency as well as maintenance strategies, including organization policies; traditional quality; and maintenance attributes, measures, and components of manufacturing performance

  ii. Pretesting and validation of the questionnaire on the representative sample of industries

  iii. Data collection using detailed manufacturing competency questionnaires through postal mail, e-mail, personal visits, interviews, and other communication means

  iv. Summarizing and analyzing the data to investigate status of various traditional maintenance strategies, besides evaluating exploits of Indian entrepreneurs with proactive competency initiatives and strategy initiatives, thus evaluating the benefits accrued in terms of manufacturing performance achievements

  v. Statistical analysis pertaining to status of various performance indicators as a result of implementations and suitable qualitative and quantitative techniques have been employed to analyze the contributions of various competency and strategy implementation dimensions towards realization of performance achievements in North Indian automobile manufacturing organizations

vi. Identification of stumbling blocks for successful competency-strategy implementation in North Indian automobile manufacturing enterprises

vii. Identification of key success factors for manufacturing competency in North Indian manufacturing automobile units.

(c) **Assessing the Actor's Capability**

The survey has been followed by case studies in selected Indian manufacturing organizations to ascertain the manufacturing performance exploits by Indian entrepreneurs. The case studies emphasize upon step-by-step implementation procedures adopted by the organizations towards achieving strategic success through manufacturing competencies. The case studies have been developed to determine the tools and techniques adopted by manufacturing organizations towards ensuring effectiveness of competency and results accrued through successful strategies. The data thus obtained regarding key performance indicators has been analyzed for arriving at their role in improving effectiveness of quality and maintenance function in the organizations. The detailed case studies include an overview of manufacturing organizations; manufacturing competency factors; strategies adopted by organizations; and their time frame, sequence, and performance enhancements accrued by the organizations.

(d) **Evolving the Management Process**

Finally, the inferences drawn from the literature review, survey (analysis of the questionnaire), and case studies have been effectively deployed to design the 'competency-strategic model' for Indian manufacturing organizations. Moreover, the guidelines for strategic measures for overcoming the barriers in implementation programs have also been evolved in the research.

3.3.3. Research Methodology

Figure 3.1 shows a block diagram for the methodology used for this research. This research has been conducted in automobile and auto parts manufacturing units in the Northern part of the country for studying the impact of these competency factors on business performance. The aim is to describe the effect of competency factors on the performance parameters of automobile manufacturing units. During this research, a large number of automobile units have been surveyed for establishing the effects of competency drivers in strategy making. A survey of various organizations has been conducted through a specially prepared questionnaire.

The approach has been directed towards analysing the effect of competency on strategies, and thus in improving firm's performance. For completion of the survey effectively, the questionnaire was prepared through

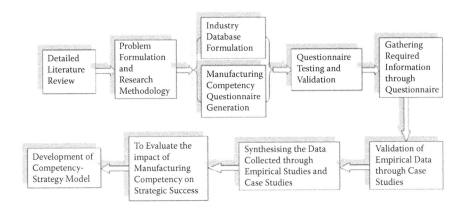

FIGURE 3.1
Block Diagram of Methodology.

extensive literature review (Ahmad and Schroeder 2011; Alsudiri et al. 2013; Amoako-Gyampah et al., 2008; Armstrong 2013; Bonjour and Micaelli 2010; Chaiprasit and Swierczek, 2011; Masoud 2013; Fernández-Pérez et al., 2012; Fernández-Mesa et al., 2013; Haartman, 2012) and was validated through vast peer review from consultants in the industry, as well as academics. The questionnaire is based on four-point Likert scale. Each performance parameter and dimension is taken as a group of several related items.

An industrial database among the automobile manufacturing organizations across the northern region of the country was created for the purpose of conducting a survey of the manufacturing competency questionnaires. The questionnaires were forwarded to the organizations and they were subsequently contacted through various communication means like postal mail, e-mail, telephonic interviews, or personal interviews through visits to various manufacturing units to describe the purpose of the research, its relevance, and to clarify any doubts or queries to facilitate responses to the manufacturing competency questionnaires.

Finalized manufacturing competency questionnaires were forwarded to approximately 350 industries that manufacture automobiles and their parts. Around 150 calls were made to interact with the persons within the industry and about 250 e-mails containing questionnaires were forwarded to various automobile units across the northern region of the country. Along with this, interviews with the appropriate persons were made and clarifications were sorted.

Furthermore, in the case of organizations with multiple products, a response for individual products were received. The responses were compiled and analyzed to determine the performance of the North Indian automobile manufacturing industry. Most respondents to the manufacturing competency questionnaires were from the top levels of management that included vice presidents, general managers (GMs), heads of operations, heads of maintenance, heads of process engineering, heads of quality assurance, quality managers, etc.

In response to all these efforts, 118 filled questionnaires were received. The simple, comprehensive, and relevant questionnaires, covering different aspects of competency and strategy factors, gathered the data required for attaining objectives of the research. A detailed description of the manufacturing competency questionnaire has been presented in Appendix I.

The various sections of the questionnaire are:

I. *Manufacturing competency*

1. Product concept (idea generation)

 1.1 Creativity

 1.2 Innovation

 1.3 Invention

 1.4 Evolution

2. Product design and development

 2.1 CAD (Technology)

 2.2 Product life cycle

 2.3 FEM/FEA

 2.4 Simulation and modeling

 2.5 Aesthetics

 2.6 Ergonomics

 2.7 Technical specifications

3. Process planning

 3.1 CAPP

 3.2 Machine selection

 3.3 Material selection

 3.4 Statistical process control

 3.5 Demand order information

4. Raw material and equipment

 4.1 Material availability

 4.2 Import

 4.3 Inventory

 4.4 Warehousing

4.5 Transportation

4.6 Automated Equipment

5. Production planning and control

5.1 CAM

5.2 Precision knowledge

5.3 Green manufacturing

5.4 System integration

5.5 Robotics

5.6 Hydraulics and pneumatics

5.7 Assembly

5.8 Finishing

5.9 Process control

6. Quality control

6.1 Inspection

6.2 Product testing

6.3 Life cycle analysis

II. *Strategic success factors*

1. Strategic agility

1.1 Price

1.2 Profit

1.3 Market share

1.4 Customer base

2. Management

2.1 Planning

2.2 Control

2.3 Project management

2.4 Information management

a) Information strategy

b) Information analysis

c) System coordination

2.5 Risk management

2.6 Performance management

2.7 Crisis management

3. Teamwork

3.1 Leadership:

a) Focus:

  i. Impact

  ii. Motivation

b) Knowledge:

  i. Conceptual

  ii. Analytical

  iii. Strategic

  iv. Expertise

c) Manage

  i. Change

  ii. Performance

3.2 Communication:

  a) Marketing

  b) Promotion

  c) Public relations

  d) Internal corporate

3.3 Cooperation

3.4 Knowledge:

  a) Knowledge transfer methodology

  b) Technology transfer

  c) International collaborations

4. <u>Administration</u>

4.1 Management

4.2 Policy formation

4.3 Management control

5. <u>Interpersonal</u>

5.1 Awareness

5.2 Self-confidence

5.3 Flexibility

5.4 Stress management

5.5 Influence

5.6 Logistics:

  a) Import

  b) Export

  c) Warehousing

  d) Waste management

  e) Product standardization

III. *Output factors*

1. Sales
2. Profit
3. Competitiveness
4. Growth and expansion
5. Production capacity
6. Production time
7. Lead time
8. Productivity
9. Market share
10. Quality
11. Reliability
12. Customer base

Figure 3.2 depicts the proposed model showing the competency factors and performance attributes for evaluating the relations between competency and strategy.

For establishing the benefits realized by an effective manufacturing competency approach, it becomes important that the effect of competency approach on different strategic success factors, and thus organization performance can be analyzed carefully.

Various statistical tools like response analysis, Cronbach's alpha, percent point scores, multiple regression analysis, ANOVA, t-tests, and Pearson correlation coefficients have been employed to evaluate and validate the contributions of competency initiatives towards building firm performance and realization of core competencies in the manufacturing organizations.

The above work has been extended by applying various qualitative techniques like AHP, TOPSIS, VIKOR, and fuzzy logic. This further has been validated by applying structural equation modeling to the above study. This study uses the confirmatory factor analysis (CFA) approach using structural equation modeling (SEM) in Analysis of Moment Structures (AMOS) 21.0 software to deploy the interrelation between competency and strategic success variables involved in the study. The data for the study have been collected through the manufacturing competency questionnaires from various North Indian automobile manufacturing industries.

Further, for assessing the actor's capability, the multiple-descriptive case study method has been used in the research and the survey has been followed by case studies in selected manufacturing organizations. In the study, the case study method has been preferred due to the following reasons:

i. The case study method implements a deeper study of an organization than data points covered under the survey.

ii. For proper research, it was needed to use multiple sources of evidence, as work cannot implement a single data collection method.

iii. Distinctive strategies are needed for research design and analysis.

While selecting the organizations for detailed case studies, the following factors have been considered:

i. The selected set of organizations should represent the manufacturing sector in terms of competition, complexity, and other aspects need to be included in the study.

ii. The selected set should include organizations, which have different manufacturing sectors.

iii. The organizations participating in the survey through the manufacturing competency questionnaire responses have been given preference.

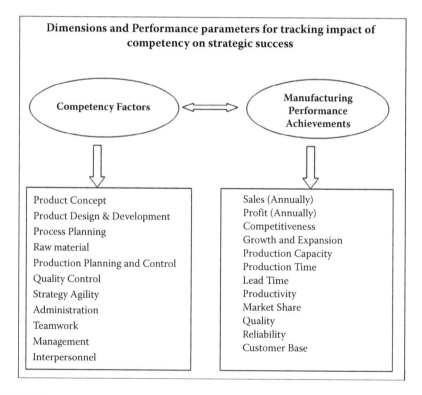

FIGURE 3.2
Competency Factors and Manufacturing Performance Achievements.

iv. There is the feasibility of getting authentic information and data related to competency from the units through personal interactions, observations, and published data. Here, it is pertinent to mention that although reasonably high numbers of questionnaire responses have been obtained from leading Indian entrepreneurs, very few organizations have come out openly and shared their exploits and performance achievements.

v. The descriptive case studies have emphasized step-by-step implementation procedure adopted by the organizations towards achieving organizational objectives. Industrial support was sought from various manufacturing organizations regarding the proposed research work, and an encouraging response has been received from the industry. The industries confirming to support the proposed research work and which were selected for doing case studies were Honda, Gurgaon (two-wheeler sector), Suzuki, Manesar (diesel and petrol cars), Mahindra and Mahindra, Chandigarh (tractors), SML Isuzu, and Roopnagar (buses and trucks).

vi. The case studies have elaborated organizational information; the need for implementation; strategies adopted; and their time frame, sequence, and impact of implementation strategies towards realization of improvements in each firm's performance.

Considering the extensive literature review, questionnaire survey, quantitative (as well as qualitative) analyses, and case studies, a competency-strategy model for the North Indian automobile manufacturing industry is developed by the authors in chapter 9. A summary of the research accomplishments has also been highlighted. Finally, the limitations of the research have been presented and recommendations for future research directions have also been suggested.

4

Reliability Analysis of Competency and Strategy

In this chapter, the impact of manufacturing competencies on the strategic success of the automobile manufacturing industry has been presented through *data analysis* and *results interpretation*. This chapter describes the analysis performed to attain the desired objectives of the study. The Statistical Package for the Social Sciences (SPSS) 21.0 (now called PASW – Predictive Analytics Software) has been used here and the following statistical techniques of this package have been applied: multiple regression, ANOVA, two tailed t-test, Cronbach alpha, and correlation.

4.1. Cronbach Alpha Reliability Analysis

The internal consistency of how closely a set of grouped variables are related in a questionnaire is measured by employing Cronbach alpha analysis. The higher the value of this coefficient, the more the generated questionnaire is reliable. Nunnaly (1978) has specified 0.7 as an acceptable reliability coefficient, but sometimes lower coefficients are also used. The reliability index is evaluated for different sections of the questionnaire which are: manufacturing competencies, strategic success, output, and the total for the overall questionnaire. Moreover, the Cronbach alpha indices are evaluated for all parameters of *manufacturing competencies, strategic success, output,* and the *overall questionnaire.*

Table 4.1a shows that the indices for manufacturing competency factors are above a value of 0.760, which reflects the internal consistency of the data response available. The indices for the various parameters of the manufacturing competencies were on the higher side – i.e., process planning (0.886), product design and development (0.879), quality control (0.828), product concept (0.826), production planning and control (0.778), and raw material and equipment (0.769) – suggesting that items have a relatively high internal consistency. The indices evaluated for the manufacturing competencies section is 0.968.

Table 4.1b shows the indices for strategic success factors. The inference drawn is that all the indices for strategic success parameters are above 0.730, which reflects the internal consistency of the data response available.

TABLE 4.1

Cronbach Alpha Reliability Index of the Questionnaire

(a) Manufacturing Competency Attributes

Product Concept	0.826
Product Design and Development	0.879
Process Planning	0.886
Raw Material and Equipment	0.769
Production Planning	0.778
Quality Control	0.828
Manufacturing Competency Section	0.968

(b) Strategic Success Attributes

Strategic Agility	0.818
Management	0.901
Teamwork	0.890
Administration	0.738
Interpersonal	0.885
Strategic Success Section	0.967

(c) Output Factors and Overall Questionnaire

Output Factors	0.906
Overall Questionnaire	0.985

The indices for the various parameters of the strategic success are on the higher side – i.e., strategic agility (0.818), management (0.901), teamwork (0.890), administration (0.738), and interpersonal (0.885) – suggesting that the items have a relatively high internal consistency. The indices evaluated for the strategic success section is 0.967.

Based on the analysis, an inference is drawn that all indices are above 0.900 – i.e., for the output section (0.906) and for the overall questionnaire used in the study (0.985) – which reflects the internal consistency of the data response available. This suggests that the items have a relatively high internal consistency.

4.2. Response Analysis

The surveyed respondents were assessed on various statements based on the parameters of the manufacturing competencies. The data was collected from the respondents on a four-point scale – i.e., not at all (A), to some extent (B), reasonably well (C), and to a great extent (D) – regarding implementation.

4.2.1. Manufacturing Competency

Different factors in Manufacturing Competency are:

1. Product Concept
2. Product Design and Development
3. Process Planning
4. Raw Material and Equipment
5. Production Planning and Control
6. Quality Control

4.2.1.1. Product Concept

Table 4.2 shows the data regarding product concept issues. The analysis of significant attributes of major product concept (idea generation) issues reveal that a significantly large number of organizations have evolved a well-planned and structured concept generation process (PPS=70.0), promote innovation and marketing (PPS=58.4), encourage inter departmental teams (PPS=51.0), centralize planning (PPS=51.0), and promote creativity (PPS=58.0). For other factors with low PPS, some improvements can be suggested. Figure 4.1 depicts the issue-wise performance.

The response analysis results showed that maximum weightage was given to the product concept attribute 'well-planned concept generation processes.' It was followed by 'company policies towards innovation,' and 'marketing department motivation for new concepts.' Almost similar preferences were there for 'centralized planning structure,' 'developments of inter departmental relationships,' and 'flexibility of organization towards changes for satisfying customers.' The analysis assessed that 23.7% of surveyed respondents reported implementing 'well planned structured concept generation process,' while 33.9% and 37.3% reported it either to some extent or reasonably well, respectively.

The product concept ideas 'innovation,' and 'marketing department's motivation for new concepts,' were also reported on the similar scale in the organization. It was analyzed that 33.9% and 45.8% reported that 'company policies promoted innovation' either to some extent or reasonably well, respectively. 48.3% and 31.4% reported that the 'marketing department was motivated enough to bring up new concepts' either to some extent or reasonably well, respectively.

When further analyzed, it was evident that all other product concepts – i.e., 'centralized planning development,' 'flexibility of organization towards changes,' and 'developing inter departmental relations for new ideas' – were

TABLE 4.2

Response Analysis of Product Concept

S. No	ISSUES	Companies Response Score				Total Responses (N)	Total Points (TPS)	Percent Points (PPS) TPS*100/ 4 * N	Central Tendency TPS/N
		A 1	B 2	C 3	D 4				
1	Do you have a well-planned and structured concept generation process in your organization?	6	40	44	28	118	330	70	2.79
2	Do your company policies promote innovation?	16	54	40	8	118	276	58.4	2.34
3	Do you feel that the marketing department is adequately motivated to get an idea about the new product?	20	37	57	4	118	281	59.5	2.38
4	Does your organization encourage the deployment of inter departmental teams to identify and create new ideas?	30	55	31	2	118	241	51.0	2.04
5	Is your organization flexible enough for making changes during operations and maintenance to satisfy customer needs?	23	74	12	9	118	243	51.5	2.06
6	Does your organization use a centralized planning structure for idea generation?	32	55	25	6	118	241	51.0	2.04
(Total Point Scored 'TPS'=A × 1 + B × 2 + C × 3 + D × 4)								56.9	2.28

up to some extent in the organizations, as the analysis reported them as 46.6%, 62.7%, and 46.6% on these product concepts. On these issues, 27.1%, 19.5%, and 25.4% of the respondents also reported that these product concepts were not implemented in their organizations.

4.2.1.2. Product Design and Development

Table 4.3 represents the performance regarding product design and development issues. The analysis of issues related to the manufacturing organization reveals that a significantly large number of organizations have an

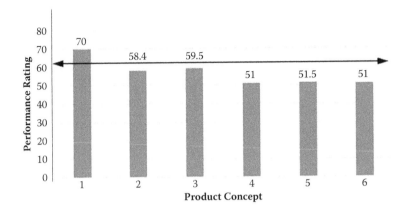

FIGURE 4.1
Performance Chart for Product Concept.

effective design technology (PPS=67.8), computer technology for analysis (PPS=54.0), product life cycle (PPS=5.2), aesthetics and ergonomics of products (PPS=57.8), and simulation and modeling (PPS=53.8).

The results showed that maximum emphasis was given to 'implementation of design technology program' in the organization, followed by 'modeling and simulating for product analysis' and 'tracking design and development costs.' Almost similar emphasis was given to 'use of product life cycles,' and 'ergonomics and aesthetics in product design,' while the least emphasis was given to 'using the computer for analyses.' Figure 4.2 represents the performance of various organizations.

The analysis assessed that 30.5% respondents reported that the 'effective design technology program' was implemented to a great extent in the organization, whereas 21.2% and 31.4% reported it either 'not at all' or 'reasonably well,' respectively. The product design and development ideas like 'use of product life cycles,' and 'usage of computerized technology for analyses', were also reported on the similar scale in the organization. It was analyzed that 63.6% and 66.1% of respondents reported their implementation 'to some extent,' while 19.5% and 10.2% reported 'no implementation,' whereas 13.6% and 21.2% reported it as 'reasonably well.' The similar trend was evident in the process of 'usage of modeling and simulation for analyzing designs,' as 50.0% respondents reported it 'to some extent', while 23.7% of the respondents reported it 'to be reasonably well,' and 20.3% reported 'not at all' in their organizations. 43.2% and 40.7% of the surveyed respondents reported 'tracking design and development costs' and 'inclusion of ergonomics and aesthetics in product designing' in their organization was at a reasonable level, while 39.0% and 32.2% respectively reported 'to some extent.'

TABLE 4.3

Response Analysis of Product Design and Development

S No.	ISSUES	Companies Response Score				Total Responses (N)	Total Points (TPS)	Percent points (PPS) TPS*100/ 4*N	Central Tendency TPS/N
		A	B	C	D				
		1	2	3	4				
1	Does your organization have an effective design technology program (CAD)?	25	20	37	36	118	320	67.8	2.71
2	Does your organization use computerized technology for analysis purposes?	12	78	25	3	118	255	54.0	2.16
3	Does the design program include procedures like product life cycle?	23	75	16	4	118	237	50.2	2.00
4	Does the design program include aesthetics and ergonomics of the product?	25	38	48	7	118	273	57.8	2.31
5	Does your organization use simulation and modeling for analyzing designs?	24	59	28	7	118	254	53.8	2.15
6	Does your organization track design and development program costs?	8	46	51	13	118	305	64.5	2.58
7	What percentage of the designing is done with the aid of computer?	45	23	31	19	118	260	55.1	2.20
(Total Point Scored 'TPS'=A × 1 + B × 2 + C × 3 + D × 4)								57.6	2.30

On the issue regarding 'higher usage of computers in designing,' it was evident that 16.3% were using for more than 75.0% of processes, 26.3% reported computer usage up to 50.0–75.0%, 19.5% reported it to in the range of 25.0–50.0%, while 38.1% of the organizations use it for less than 25.0%.

4.2.1.3. Process Planning

Table 4.4 portrays the data regarding process planning issues. An analysis of process planning issues shows that most organizations have generally

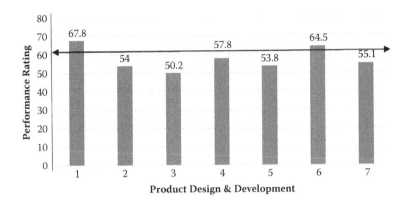

FIGURE 4.2
Performance Chart for Product Design and Development.

scored a low rating. The data shows that many manufacturing organizations have an effective process planning program (PPS = 67.16), tracking process planning costs (PPS = 70.70), material and machine selection (PPS = 56.10), group technology (PPS = 5.80), and the finishing and assembly of the product (PPS = 63.70).

The idea of 'implementation of a design technology program' was given maximum preference which was followed by 'taking and simulating account of assembling and finishing of products' and 'tracking process planning costs.' Similar emphasis was given to 'preferences to departments integration' and 'usage of mechanism for machine and material selection,' while the least preference was for 'software-based planning and the regular updating of software.'

Evidently, regarding 'higher usage of computerized process planning,' 7.6% of organizations were using it for more than 75.0%, 18.6% of organizations reported the use of computers up to 50.0–75.0%, 38.1% reported it to be in range of 25.0–50.0% and 35.6% of the organizations were using them less than 25.0%. The analysis showed that 27.1% of respondents reported an 'effective process planning program' to a great extent in their organization, whereas 24.7% and 38.1% reported it either to 'reasonably well' or 'to some extent,' respectively. 20.3% reported the 'usage of mechanism for machine and material selection,' as 'reasonably well' while 71.2% reported 'to some extent'. Figure 4.3 represents the performance of various organizations.

Regarding the issues based on process planning – i.e. 'preferences to departments integration,' 'use of group technology,' and 'planning software regularly updated and reviewed with technological advancement' – 20.3%, 24.7%, and 23.7% of the organizations reported that there

TABLE 4.4

Response Analysis of Process Planning

S No.	FACTORS	No. of Companies Scoring Points A 1	B 2	C 3	D 4	Total No. of Responses (N)	Total Points Scored (TPS)	Percent points Scored (PPS) TPS 100/4*N	Central Tendency TPS/N
1	Does your organization have an effective process planning program?	12	45	29	32	118	317	67.2	2.69
2	Does your organization apply group technology?	29	59	27	3	118	240	50.8	2.03
3	Does your organization possess a mechanism for material and machine selection?	5	84	24	5	118	265	056.1	2.25
4	Is the planning software is updated and reviewed periodically in accordance with technological changes?	28	56	31	3	118	245	51.9	2.08
5	Does your organization track process planning costs?	11	38	69	10	118	334	70.7	2.83
6	Does your organization prefer the integration of different departments?	24	51	43	0	118	255	54.0	2.16
7	Does your organization take into account the finishing and assembly of the product?	4	57	45	12	118	301	63.7	2.55
8	What percentage of the process planning is done with the aid of technology?	42	45	22	9	118	234	49.6	1.98
(Total Point Scored 'TPS' = A × 1 + B × 2 + C × 3 + D × 4)								58.0	2.32

was no implementation at all; whereas 43.2%, 50.0%, and 47.5% reported their implementation was there 'to some extent'. On the issues of 'process planning costs,' and 'considering assembling and finishing of products,' 50.0% and 39.0% of the organizations reported reasonably well implementation, while 32.2% and 48.3% reported the implementation 'to some extent'.

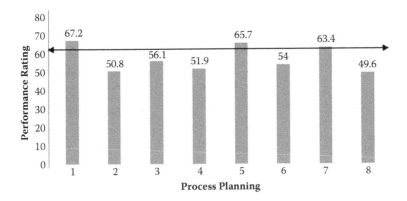

FIGURE 4.3
Performance Chart for Process Planning.

4.2.1.4. Raw Material and Equipment

Table 4.5 illustrates performance regarding the raw material and equipment issues.

The analysis of various issues related to manufacturing organizations indicated that many manufacturing organizations use ERP software for record keeping (PPS=68.0), whereas some other factors – like having own their transportation (PPS=50.2), inventory storage (PPS=58.2), and different departments involved in machine selection (PPS=55.7) – need immediate attention, since these factors have been found to be underperforming. Figure 4.4 illustrates the issue-wise performance.

The results show that the 'use of ERP software for record keeping' was given the maximum emphasis, followed by the 'existence of warehouse facilities for inventory, synergistically involving the designing, production, and marketing department in equipment selection' and the 'availability of automated equipment having process capabilities equivalent to market demands.' The least emphasis was given to the 'existence of a transportation facility.'

The 'use of ERP software for record keeping' was reported by 28.8% of the organizations 'to a great extent' while 29.7% and 26.3% reported it either as 'reasonably well' or 'to some extent,' respectively. On the issue regarding 'existence of a transportation facility,' 41.5% did not have one at all, while 23.7% and 27.1% reported it either 'to some extent' or 'reasonably well,' respectively. Around 21.2% organizations reported that they had no 'warehouse for inventory storages,' while 27.1% and 38.1% reported it either 'at a reasonable level' or 'to some extent,' respectively. 37.3% and 33.9% reported that 'availability of automated equipment having process capabilities,' and 'synergistic involvement of designing, production, and

TABLE 4.5

Response Analysis of Raw Material and Equipment

S No.	ISSUES	Companies Response score				Total Responses (N)	Total points (TPS)	Percent points (PPS) PS*100 4*N	Central Tendency TPS/N
		A	B	C	D				
		1	2	3	4				
1	Does your organization use ERP software for record keeping?	18	31	35	34	118	321	68	2.72
2	Does your organization have its own transportation?	49	28	32	9	118	237	50.2	2.01
3	Does your organization have enough ware-houses for inventory storage?	25	45	32	16	118	275	58.2	2.33
4	Are the three depart-ments (marketing, designing, and produc-tion) synergistically involved in equipment selection decisions?	19	56	40	3	118	263	55.7	2.23
5	Does your organization have sufficient auto-mated equipment with approximate process capabilities to meet market demands?	14	58	44	2	118	270	57.2	2.29
(Total Point Scored 'TPS' = A × 1 + B × 2 + C × 3 + D × 4)								57.86	2.32

marketing department in equipment selection,' was 'at a reasonable level' while 49.2% and 47.5% organization reported 'to some extent.'

4.2.1.5. Production Planning and Control

Table 4.6 reports performance regarding production planning and control issues.

The analysis of data obtained from the survey has indicated that many manufacturing organizations have generally reported low performance regarding production, planning, and control factors. The results reveal that many manufacturing organizations prioritize precision and accuracy (PPS = 65.9), green manufacturing (PPS = 52.1), hydraulic and pneumatic systems (PPS = 5.0), and computer-aided manufacturing (PPS-59.1). Figure 4.5 shows the chart for issue-wise performance.

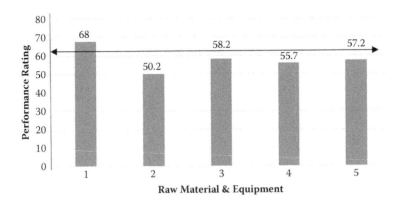

FIGURE 4.4
Performance Chart for Raw Material and Equipment.

The results show that 'precise and accurate results' was given the highest emphasis, followed by the 'computerized manufacturing systems,' and 'tracking production planning and control programs cost.' The minimum emphasis was on the issue of 'green manufacturing.'

The analysis for the issues 'manufacturing hours in relation to working hours' and 'help of robots,' revealed that 76.3% and 77.1% reported these for less than 25.0% of the time in organizations while 22.9% and 21.2% reported these for 25.0–50.0% of the time. Similarly, 43.2% of organizations reported the use of 'hydraulic and pneumatic systems' for 25.0–50.0%, while 33.9% reported for less than 25.0%. Around 41.5% organizations used 'computerized manufacturing systems' to some extent, while 34.8% and 15.3% were at either a 'reasonable level' or 'not at all,' respectively. About 44.1% and 46.6% of the organizations followed 'tracking production planning and control program costs,' and 'green manufacturing' to some extent, while 11.0% and 27.1% stated 'not at all.' Around 30.5% of organizations 'exert to get precise and accurate results' to some extent while 57.6% were at a reasonable level.

4.2.1.6. Quality Control

Table 4.7 outlines performance regarding quality control issues.

The analysis of significant attributes of quality reveals that most manufacturing organizations have shown an acceptable performance rating. The data has indicated that many manufacturing organizations test products under actual conditions (PPS = 75.4), use technology for quality analysis (PPS = 57.0), and carry-out life cycle analysis (PPS = 52.7). Figure 4.6 outlines the issue-wise performance.

TABLE 4.6

Response Analysis of Production Planning and Control

S No.	ISSUES	Companies Response Score				Total Responses (N)	Total points (TPS)	Percent points (PPS) TPS*100/ 4*N	Central Tendency TPS/N
		A	B	C	D				
		1	2	3	4				
1	Does your organization have a computerized manufacturing systems (CAM)?	18	49	41	10	118	279	59.1	2.36
2	How much do you exert to get precise and accurate dimensions?	7	36	68	7	118	311	65.9	2.64
3	Does your organization prefer green manufacturing?	32	55	20	11	118	246	52.1	2.08
4	Does your organization track production planning and control program costs?	13	52	42	11	118	287	60.8	2.43
5	What percentage of the work is done with the help of robots?	91	25	0	2	118	149	31.5	1.26
6	What is the percentage of maintenance hours in relation to total working hours?	90	27	1	0	118	147	31.1	1.25
7	To what extent are hydraulic and pneumatic systems employed in your organization?	40	51	14	13	118	236	50.0	2.00
(Total Point Scored 'TPS' = A × 1 × B × 2 + C × 3 + D × 4)								50.07	1.86

The response analysis results show that 'testing product materials under actual conditions' was given the maximum preference followed by 'technology to analyze quality,' and 'life cycle analysis of products.' The least emphasis was given to the issue of 'computerized quality control instructions.' The analysis shows that in the 'reprocessing of products after inspection,' 72.9% reported it less than 25.0%, while 23.7% reported it 25.0–50.0%.

On the contrary, under 'investment on quality control and inspection in comparison to total production cost,' 43.2% reported it less than 25.0% while 35.6% reported as 25.0–50.0%. It was further analyzed that regarding the quality control parameters like 'use of technology for the assessment of the products,' 'computerized quality control instructions,' and 'life cycle analysis

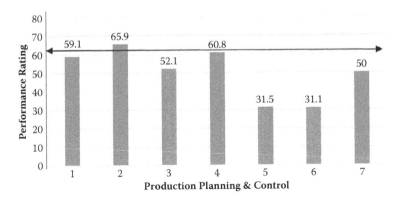

FIGURE 4.5
Performance Chart for Production Planning and Control.

TABLE 4.7

Response Analysis of Quality Control

| S No. | ISSUES | Companies Response Score | | | | Total Responses (N) | Total points (TPS) | Percent points (PPS) TPS*100/ 4*N | Central Tendency TPS/N |
| | | A | B | C | D | | | | |
		1	2	3	4				
1	Does your organization, test products under actual conditions?	8	14	64	32	118	356	75.4	3.02
2	Does your organization carry out life cycle analysis of the product?	28	61	17	12	118	249	52.7	2.11
3	Does your organization use technology to analyze quality?	30	37	39	12	118	269	57.0	2.28
4	Does your organization issue computerized quality control instructions?	41	53	24	0	118	219	46.4	1.86
5	Up to what extent does the product need to be reprocessed after inspection?	86	28	2	2	118	156	33	1.32
6	To what extent does your organization, invest in quality control and inspection when compared to the total production cost?	51	42	23	2	118	212	44.9	1.80
(Total Point Scored 'TPS' = A × 1 + B × 2 + C × 3 + D × 4)								51.57	2.07

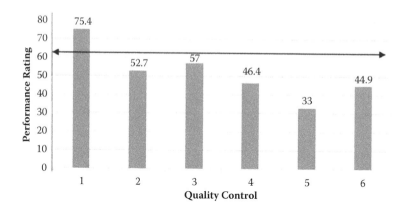

FIGURE 4.6
Performance Chart for Quality Control.

of the products,' 25.4%, 34.8%, and 23.7% were not following them at all, whereas on same issues 51.7%, 31.4%, and 44.10% were using them to some degree. It was inferenced that 54.2% of organizations were 'testing products under actual conditions' at a reasonable level while 27.1% to a great extent.

4.2.2. Strategic Success

Various parameters of strategic success are:

1. Strategic Agility
2. Management
3. Teamwork
4. Administration
5. Interpersonal

4.2.2.1. Strategic Agility

Table 4.8 depicts performance on Strategic Agility issues.

The results reveal that many manufacturing organizations have exhibited a strong endeavor to realize various strategic success issues, such as quality conformance (PPS=66.9), a strong customer base (PPS=62.3), competitive pricing (PPS=65.0), market share (PPS=71.4), and profit (PPS=64.0). The data reveals that many organizations press for various strategic issues like a strong customer base, quality performance, competitive pricing, market share, and profit. Figure 4.7 depicts the issue-wise

TABLE 4.8

Response Analysis of Strategic Agility

S No.	ISSUES	A 1	B 2	C 3	D 4	Total Responses (N)	Total points (TPS)	Percent points (PPS) TPS*100/ 4*N	Central Tendency TPS/N
		Companies Response Score							
1	Quality conformance	10	33	60	15	118	316	66.9	2.68
2	Improving customer base	9	47	57	5	118	294	62.3	2.49
3	Developing and enhancing market share	11	45	42	20	118	307	65.0	2.60
4	Achieving a higher profit	7	33	48	30	118	337	71.4	2.86
5	Competitive pricing of products	15	48	29	26	118	302	64.0	2.56
(Total Point Scored 'TPS'=Ax1+Bx2+Cx3+Dx4)								**65.92**	**2.64**

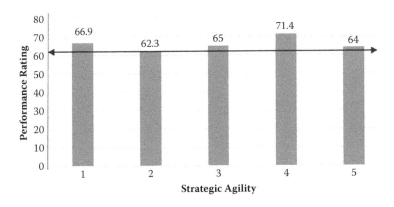

FIGURE 4.7
Performance Chart for Strategic Agility.

performance. The results showed that under strategic agility, 'achieving a higher profit' was given maximum emphasis over 'quality performance' and 'market share.' A similar preference was given to 'competitive pricing of the products,' while the least was given to 'improving customer base.'

'Quality performance' and 'customer base,' 50.8% and 48.3% respectively, responded these to be followed reasonably well, whereas 28.0% and 39.8% reported them to some extent. On similar pattern, 40.7% of the organizations followed 'achieving higher profits,' at reasonable level, while 28.4% achieved it at some level and 25.4% of organizations at a high level. On the issue of 'market share,' 35.6% and 38.1% implemented it at either reasonable level or to some extent, while 16.9% organizations implemented it to a great extent. 'Competitive prices' were also achieved to some extent in 40.7%, while 22.0% and 24.7% were working on this either to a great extent or at a reasonable level.

4.2.2.2. Management

Table 4.9 portrays performance regarding the issues on management.

The significant attributes of management depict that many manufacturing organizations have better production planning and control functions (PPS=72.4), coordination between departments (PPS=7.0) and enhanced production capabilities (PPS=71.2), but some improvement

TABLE 4.9

Response Analysis of Management

S No.	ISSUES	Companies Response Score				Total Responses (N)	Total points (TPS)	Percent points (PPS) TPS*100/ 4*N	Central Tendency TPS/N
		A	B	C	D				
		1	2	3	4				
1	Enhanced production capabilities and improved control	3	39	49	27	118	336	71.2	2.85
2	Better production planning and control functions	2	35	54	27	118	342	72.4	2.89
3	Information flow within departments through the Internet	37	24	46	11	118	267	56.6	2.26
4	Information analysis in different departments	32	34	51	1	118	257	54.4	2.18
5	Risk management	33	49	33	3	118	242	51.3	2.05
6	Crisis management	25	49	44	0	118	255	54.0	2.16
7	Coordination between departments	6	36	52	24	118	330	70.0	2.80
(Total Point Scored 'TPS' = A × 1 + B × 2 + C × 3 + D × 4)								61.41	2.46

can be suggested for other factors like efficient information flow (PPS=56.6), information analysis in departments (PPS=54.4), crisis management (PPS=54.0), and risk management (PPS=51.3), since these factors have been underperforming.

The results show that 'production planning and control functions' were given maximum preference, followed by 'coordination between departments' and 'production capabilities and improved control.' The 'information flow within the departments through the Internet' was also given similar preferences, whereas equal importance was for 'information analysis in various departments' and 'crisis management.' The least preference was given to 'risk management.' Figure 4.8 depicts the issue-wise performance.

The analysis shows that for the issues 'production capabilities and improved control,' 'production planning and control functions,' 'coordination between departments,' and 'information analysis in various departments,' 41–46% of organizations were realizing them at reasonable level whereas 29–33% of organizations were realizing them to some extent, respectively. Also, it was seen that 20–22% of organizations were realizing these issues to a great extent.

It was further analyzed that 41.5% of the organizations were working on the concept of 'crisis management' and 'risk management' to some extent, while 28.0% and 21.2% were not implementing them, respectively, and 28.0% and 37.3% were implementing them at reasonable level. Around 39.0% of the organizations reported 'information flow in the departments through the internet' at a reasonable level while 20.3% reported this to some extent and 31.4% of the organizations were not using this concept at all.

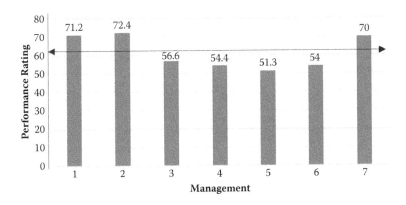

FIGURE 4.8
Performance Chart for Management.

4.2.2.3. Teamwork

Table 4.10 represents performance regarding the teamwork issues.

The results revealed that many manufacturing organizations emphasize on improving communication and coordination amongst team members (PPS=68.8), coordinated efforts for next generation technology (PPS=62.1), and better promotion of products (PPS=68.2), whereas some other factors need immediate attention, since these factors have been found to be underperforming. The results show that 'cooperation and communication between team members' and 'promoting developed products' were given the highest emphasis, followed by 'coordinate efforts for next generation technology' and 'effective management of process capabilities.' Figure 4.9 represents the issue-wise performance.

TABLE 4.10

Response Analysis of Teamwork

S No.	ISSUES	Companies Response Score				Total Responses (N)	Total points (TPS)	Percent points (PPS) TPS*100/ 4*N	Central Tendency TPS/N
		A	B	C	D				
		1	2	3	4				
1	Coordinated efforts for the development/ fostering of next genera-tion technology	8	45	65	0	118	293	62.1	2.48
2	Transforming a traditional hierarchical organization into a boundary-less organization	27	68	22	1	118	233	49.3	1.97
3	Promotion of developed product	14	29	50	25	118	322	68.2	2.73
4	Culture of Kaizen and continuous improvement	37	37	26	18	118	261	55.3	2.21
5	Overall equipment effectiveness (OEE) improvement	30	41	28	19	118	272	57.6	2.31
6	Effectively managing process capability	13	57	45	3	118	274	58.0	2.32
7	Enhanced autonomous maintenance capabilities	30	48	26	14	118	260	55.1	2.20
8	Communication and cooperation among team members	8	39	45	26	118	325	68.8	2.75
(Total Point Scored 'TPS' = A × 1 + B × 2 + C × 3 + D × 4)								59.3	2.37

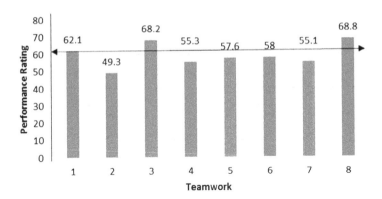

FIGURE 4.9
Performance Chart for Teamwork.

The scope of 'overall equipment effectiveness improvement' was also given preference, while equal importance was for 'Kaizen and continuous improvement' and 'enhanced autonomous manufacturing capabilities.' The least preference was given to 'transforming hierarchical organization into boundary-less organization.' It was further concluded that 23.0%–25.0% were following concepts like 'transforming hierarchical organization into boundary-less organization,' 'overall equipment effectiveness,' and 'autonomous manufacturing capabilities,' while the same issues were followed to some extent by 57.6%, 34.8%, and 40.7%, respectively.

The 'Kaizen and continuous improvement' was either not followed or followed to some extent by 31.4%, while 22.0% followed it at a reasonable level. The factors 'communication and cooperation between team members' and 'effective managing process capability' were followed at reasonable level in 31.4% of organizations, whereas 33.3% and 48.3% followed them to some extent. Around 54.2% of the organizations reported that a 'coordinated effort for development of new technology' was followed at a reasonable level while 38.1% followed it to some extent. Similarly, 42.4% of the organizations reported that a 'promotion of developed product concept' was followed at a reasonable level, while 21.2% and 24.7% organizations followed it a great extent and to some extent, respectively.

4.2.2.4. Administration

Table 4.11 illustrates the performance regarding administration issues. The analysis of significant attributes of administration has indicated that many manufacturing organizations have an efficient administration and

TABLE 4.11

Response Analysis of Administration

S No.	ISSUES	Companies Response Score				Total Responses (N)	Total points (TPS)	Percent points (PPS) TPS*100/ 4*N	Central Tendency TPS/N
		A	B	C	D				
		1	2	3	4				
1	Efficient office administration and management	5	25	69	19	118	338	71.6	2.86
2	Policy formation	8	42	45	23	118	319	67.6	2.70
3	Commitment of top-level management	7	61	31	19	118	298	63.1	2.53
4	Support and encouragement from top-level management	19	54	41	4	118	266	56.3	2.25
(Total Point Scored 'TPS' = A × 1 + B × 2 + C × 3 + D × 4)								64.65	2.59

management (PPS=71.6), support and encouragement from top management (PPS=56.3), policy formation (PPS=67.6), and top-level management commitment (PPS=63.1). Figure 4.10 illustrates the issue-wise performance.

The results show that 'office management and administration' was given the highest emphasis, followed by 'policy formation.' The 'commitment of top-level management' was also given some emphasis while the least preference was given to 'support and encouragement from top-level management.' The analysis of administrative reforms such as 'office administration and management' was followed at a reasonable level in 57.6% of organizations, while 21.2% reported it to some extent. Around 51.7% reported 'commitment from top management' to some extent, whereas 26.3% reported it at a reasonable level. It was further assessed that in 35.8% organizations 'policy formation' and 'support and encouragement from top management' was at a reasonable level, while 35.6% and 45.8% were implementing it to some extent.

4.2.2.5. Interpersonal

Table 4.12 outlines performance regarding the interpersonal factors.

The analysis of significant attributes of interpersonal issues has revealed that many manufacturing organizations have worked aggressively for building the self-confidence of employees (PPS = 71.6), safety and health awareness (PPS = 58.7), self-managed teams (PPS = 57.6), employee

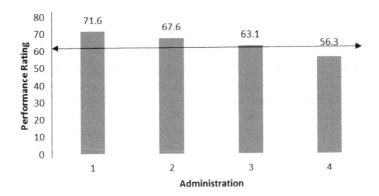

FIGURE 4.10
Performance Chart for Administration.

TABLE 4.12

Response Analysis of Interpersonal

S No.	ISSUES	Companies Response Score				Total Responses (N)	Total points (TPS)	Percent points (PPS) TPS*100/ 4*N	Central Tendency TPS/N
		A	B	C	D				
		1	2	3	4				
1	Self-confidence of employees	0	30	74	14	118	338	71.6	2.86
2	Stress management	38	43	25	12	118	247	52.3	2.09
3	Waste utilization	39	60	7	12	118	228	48.3	1.93
4	Multi skilled workers	28	39	50	1	118	260	55.0	2.20
5	Safety and health awareness among workers	17	55	34	12	118	277	58.7	2.34
6	Broader job Perspectives and employee empowerment	23	46	43	6	118	268	56.8	2.27
7	Self-managed project teams and problem solving groups	30	31	48	9	118	272	57.6	2.31
(Total Point Scored 'TPS' = A × 1 + B × 2 + C × 3 + D × 4)								57.19	2.29

empowerment (PPS = 56.8), a multi skilled workforce (PPS = 55.0) and stress management (PPS = 52.3). The results reveal that the 'self-confidence of employees' was given the highest emphasis, followed by 'safety and health awareness amid workers,' and 'self-managed and problem-solving project teams.' The scope of the 'multi skilled workers,' 'job prospective and employee empowerment,' and 'stress management' were given somewhat equal preference. The least preference was given to 'waste utilization.' Figure 4.11 outlines the issue-wise performance.

After analysis, it was assessed that 62.7% of the organizations described 'self-confidence of the employees' at a reasonable level while it was achieved 'to some extent' in 25.4% of the organizations. Approximately 33.0% of the organizations stated 'not at all' following interpersonal issues like 'stress management' and 'waste utilization,' whereas 10.2% noted a great extent on both issues. Around 36.4% of the organization reported to follow 'stress management' to some extent, whereas 21.2% of organizations followed it at a reasonable level. Similarly, 50.8% organizations were working on 'waste utilization' to some extent.

It was assessed that about 25.0% of the organizations were not implementing the concept of 'multi skilled workers,' and 'self-managed project teams and problem-solving groups' whereas about 42.0% of the organizations were implementing both these concepts at reasonable levels. It was also concluded that 33.1% of organizations were pushing the concepts of 'multi skilled workers' to some extent while 26.3% of the organizations were aiming for 'self-managed project teams and problem-solving groups.' On the issue of 'safety and health awareness among workers,' 46.6% of the organizations were following it to some extent whereas 28.8% organizations were doing it on a reasonable level. On similar notes, regarding 'broader job prospective and employee empowerment' 39.9% of the

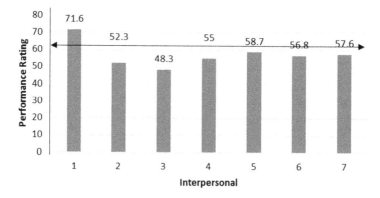

FIGURE 4.11
Performance Chart for Interpersonal.

organizations were implementing it to some extent whereas 36.4% organizations were doing it on reasonable level.

4.2.3. Output

Table 4.13 characterizes the performance of manufacturing organizations regarding output factors.

The analysis of significant attributes of organizations has revealed that many manufacturing organizations have shown acceptable performance regarding major attributes. The organizations' work on quality (PPS=79.2), production capacity (PPS=77.0), reliability (PPS=75.0), production time (PPS=73.5), productivity (PPS=68.6), competitiveness (PPS=63.7), sales (annually) (PPS=65.2), customer base (PPS=64.0), and profit (annually) (PPS=62.7).

However, there is an emergent need for improvements in lead time (PPS=57.6), growth and expansion (PPS=58.4), and market share (PPS=59.7), since these factors have been underperforming. Figure 4.12 depicts the issue-wise performance. The analysis shows the parameters like 'production time,' 'production capacity,' 'quality,' and 'reliability,' were in excellent condition in 30.0–34.0% of the organizations, while

TABLE 4.13

Response Analysis of Output

S No.	ISSUES	Companies Response Score				Total Responses (N)	Total points (TPS)	Percent points (PPS) TPS*100/4*N	Central Tendency TPS/N
		A 1	B 2	C 3	D 4				
1	Production capacity	0	30	49	39	118	363	77.0	3.07
2	Production time	5	32	46	35	118	347	73.5	2.94
3	Lead time	18	52	42	6	118	272	57.6	2.31
4	Quality	3	14	61	40	118	374	79.2	3.17
5	Reliability	6	29	42	41	118	354	75.0	3.00
6	Productivity	9	43	35	31	118	324	68.6	2.74
7	Growth and expansion	23	38	51	6	118	276	58.4	2.34
8	Competitiveness	16	26	71	5	118	301	63.7	2.55
9	Sales (annually)	5	61	27	25	118	308	65.2	2.61
10	Profit (annually)	14	57	20	27	118	296	62.7	2.51
11	Market share	7	62	45	4	118	282	59.7	2.39
12	Customer base	6	55	42	15	118	302	64.0	2.56

(Total Point Scored 'TPS' = A × 1 + B × 2 + C × 3 + D × 4)

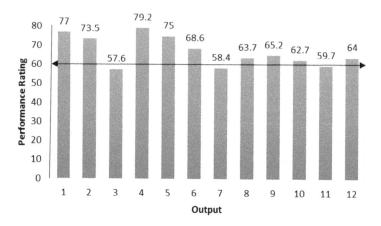

FIGURE 4.12
Performance Chart for Output Factors.

26.3% organizations had 'productivity' in excellent condition. 21.0%–23.0% of the organizations had process like 'sales' and 'profit' in excellent conditions.

On the other hand, regarding other processes, the conditions were too encouraging. 'Production capacity' was in good condition among 41.5% of the organizations while 25.4% organizations reported it as satisfactory. 'Production time' was in good condition among 39.0% of the organizations while 27.1% organization reported it as satisfactory. 'Lead time' was in good condition among 35.6% of the organizations while 44.1% organization reported it as satisfactory.

'Quality' was in good condition among 51.7% of the organizations while 11.9% of organizations reported it as satisfactory. 'Reliability' was in good condition among 35.6% of the organizations while 24.6% organization reported it as satisfactory. 'Productivity' was in good condition among 29.7% of the organizations while 36.4% organization reported it as satisfactory. 'Growth and expansion' was in good condition among 43.2% of the organizations while 32.2% organization reported it as satisfactory.

'Competitiveness' was in good condition among 60.2% of the organizations while 22.0% organization reported it as satisfactory. 'Sales (annually)' were in good condition among 22.9% of the organizations while 51.7% organization reported it as satisfactory. 'Profit (annually)' was in good condition among 16.9% of the organizations while 48.3% organization reported it as satisfactory. 'Market share' was in good condition among 38.1% of the organizations while 52.5% organization reported it as satisfactory. 'Customer base' was in good condition among 35.6% of the organizations while 46.6% organization reported it as satisfactory.

4.2.4. Comparative Analysis of All Factors

Based on the above analysis of percent point score (PPS) and central tendency, the following Table 4.14 depicts the comparative result of all factors.

4.3. Correlation Analysis

The purpose of correlation analysis is to identify the relationship within various parameters. Moreover, perception was measured by correlation

TABLE 4.14

Comparative Result of All Factors

Factors	Percent Points Score (PPS)	Central Tendency	Rank
Manufacturing Competency Factors			
Process planning	580	2.32	1
Raw material and equipment	57.86	2.32	2
Product design and development	57.60	2.30	3
Product concept	56.90	2.28	4
Quality control	51.57	27	5
Production planning and control	507	1.86	6
Strategic Success Factors			
Strategic agility	65.92	2.64	1
Administration	64.65	2.59	2
Management	61.41	2.46	3
Teamwork	59.30	2.37	4
Interpersonal	57.19	2.29	5
Output Factors			
Quality	79.20	3.17	1
Production capacity	770	37	2
Reliability	750	30	3
Production time	73.50	2.94	4
Productivity	68.60	2.74	5
Sales (annually)	65.20	2.61	6
Customer base	640	2.56	7
Competitiveness	63.70	2.55	8
Profit (annually)	62.70	2.51	9
Market share	59.70	2.39	10
Growth and expansion	58.40	2.34	11
Lead time	57.60	2.31	12

by evaluating statements as all were measured on the same scale. The correlation process used is the Karl Pearson correlation with a significance level of 0.05. Table 4.15 shows the same.

The purpose of correlation matrix was to establish the relationship and its direction between the parameters of the manufacturing competencies and strategic success over the output process followed in the organizations. The hypothesis was also framed for the relationship between the parameters at 0.05 levels of significance.

H_{01}: There was no relationship between the product concept and output process.

The analysis of the correlation matrix showed that the above null hypothesis assumed was not acceptable, as the correlations obtained between the product concept and all process of output were significant and affected the organization in a positive manner. It was inferred that the correlation of product concept with the parameters – i.e., reliability (r = 0.727), competitiveness (r = 0.684), quality (r = 0.675), production time (r = 0.613), production capacity (r = 0.606), growth and expansion (r = 0.527), and productivity (r = 0.500) – was significantly positive.

TABLE 4.15

Karl Pearson Correlation Matrix

		OUTPUTS											
		O1	O2	O3	O4	O5	O6	O7	O8	O9	O10	O11	O12
	I1	.606	.613	.450	.675	.727	.500	.527	.684	.404	.453	.448	.348
	I2	.606	.635	.527	.692	.758	.554	.556	.696	.472	.505	.471	.436
	I3	.550	.469	.371	.595	.746	.468	.472	.675	.450	.451	.354	.359
	I4	.672	.631	.451	.690	.770	.553	.552	.704	.478	.511	.526	.412
INPUTS	I5	.614	.549	.384	.590	.688	.496	.553	.638	.443	.422	.508	.481
	I6	.653	.628	.427	.586	.652	.564	.531	.525	.395	.377	.358	.453
	I7	.679	.677	.506	.709	.643	.612	.590	.569	.547	.554	.482	.474
	I8	.516	.640	.597	.547	.700	.581	.606	.639	.529	.655	.521	.515
	I9	.607	.652	.494	.647	.806	.596	.555	.708	.522	.584	.508	.467
	I10	.527	.566	.480	.624	.685	.575	.482	.631	.491	.507	.413	.488
	I11	.528	.554	.450	.553	.738	.470	.449	.610	.400	.432	.475	.498

(O1 – Production capacity, O2 – Production time, O3 – Lead time, O4 – Quality, O5 – Reliability, O6 – Productivity, O7 – Growth and expansion, O8 – Competitiveness, O9 – Sales, O10 – Profit, O11 – Market share, O12 – Customer base, I1 – Product concept, I2 – Product design and development, I3 – Process planning, I4 – Raw material and equipment, I5 – Production planning and control, I6 – Quality control, I7 – Strategic Agility, I8 – Management, I9 – Teamwork, I10 – Administration, I11 – Interpersonal)

H$_{02}$: There was no relationship between the product design and development and output process.

The analysis of the correlation matrix showed that the above null hypothesis assumed was not acceptable, as the correlations obtained between the product design and development and all process of output were significant and affected the organization in a positive manner. It was inferred that the correlation of product design and development with the parameters – i.e., reliability (r = 0.758), competitiveness (r = 0.696), quality (r = 0.692), production time (r = 0.635), production capacity (r = 0.606), growth and expansion (r = 0.556), productivity (r = 0.554), lead time (r = 0.527), and profit (r = 0.505) – were significantly positive.

H$_{03}$: There was no relationship between the process planning and output process.

The analysis of the correlation matrix showed that the above null hypothesis assumed was not acceptable, as the correlations obtained between the process planning and all process of output were significant and affected the organization in a positive manner. It was inferred that the correlation of process planning with the parameters – i.e., reliability (r = 0.746), competitiveness (r = 0.675), quality (r = 0.595), and production capacity (r = 0.550) – were significantly positive.

H$_{04}$: There was no relationship between the raw material and equipment and output process.

The analysis of the correlation matrix showed that the above null hypothesis assumed was not acceptable, as the correlations obtained between the raw material and equipment and all process of output were significant and affected the organization in a positive manner. It was inferred that the correlation of the raw material and equipment with the parameters – i.e., reliability (r = 0.770), competitiveness (r = 0.704), quality (r = 0.690), production capacity (r = 0.672), production time (r = 0.631), productivity (r = 0.553), growth and expansion (r = 0.552), market share (r = 0.526), and profit (r = 0.511) – was significantly positive.

H$_{05}$: There was no relationship between the production planning and output process.

The analysis of the correlation matrix showed that the above null hypothesis assumed was not acceptable, as the correlations obtained between the production planning and all process of output were significant and affected the organization in a positive manner. It was inferred that the correlation of production planning with the parameters – i.e., reliability (r = 0.688), competitiveness (r = 0.638), production capacity (r = 0.614), quality (r = 0.590), growth and expansion (r = 0.553), production time (r = 0.549), and market share (r = 0.508) – was significantly positive.

H$_{06}$: There was no relationship between quality control and output process.

The analysis of the correlation matrix showed that the above null hypothesis assumed was not acceptable, as the correlations obtained between quality control and all process of output were significant and affected in the organization in a positive manner. It was inferred that the correlation of the quality control with the parameters – i.e., production capacity (r = 0.653), reliability (r = 0.652), production time (r = 0.628), quality (r = 0.586), productivity (r = 0.564), growth and expansion (r = 0.531), and competitiveness (r = 0.525) – was significantly positive.

H$_{07}$: There was no relationship between the strategic agility and output process.

The analysis of the correlation matrix showed that the above null hypothesis assumed was not acceptable, as the correlations obtained between strategic agility and all process of output were significant and affected the organization in a positive manner. It was inferred that the correlation of strategic agility with the parameters – i.e., quality (r = 0.709), production capacity (r = 0.679), production time (r = 0.677), reliability (r = 0.643), productivity (r = 0.612), growth and expansion (r = 0.590), competitiveness (r = 0.569), profit (r = 0.554), sales (r = 0.547), and lead time (r = 0.506) – was significantly positive.

H$_{08}$: There was no relationship between the management and output process.

The analysis of the correlation matrix showed that the above null hypothesis assumed was not acceptable, as the correlations obtained between management and all process of output were significant and affected the organization in a positive manner. It was inferred that the correlation of the management with the parameters – i.e., reliability (r = 0.700), profit (r = 0.655), production time (r = 0.640), competitiveness (r = 0.639), growth and expansion (r = 0.606), lead time (r = 0.597), productivity (r = 0.581), quality (r = 0.547), sales (r = 0.529), market share (r = 0.521), production capacity (r = 0.516), and customer base (r = 0.515) – was significantly positive.

H$_{09}$: There was no relationship between the team work and output process.

The analysis of the correlation matrix showed that the above null hypothesis assumed was not acceptable, as the correlations obtained between the teamwork and all process of output were significant and they were being affected in the organization in a positive manner. It was inferred that the correlation of the teamwork with the parameters – i.e., reliability (r = 0.806), competitiveness (r = 0.708), production time (r = 0.652), quality (r = 0.647), production capacity (r = 0.607), productivity (r = 0.596), profit (r = 0.584), growth and expansion (r = 0.555), sales (r = 0.522), and market share (r = 0.508) – was significantly positive.

H_{10}: There was no relationship between the administration and output process.

The analysis of the correlation matrix showed that the above null hypothesis assumed was not acceptable, as the correlations obtained between the administration and all process of output were significant and affected the organization in a positive manner. It was inferred that the correlation of the administration with the parameters – i.e., reliability (r = 0.685), competitiveness (r = 0.631), quality (r = 0.624), productivity (r = 0.575), production time (r = 0.566), production capacity (r = 0.527), and profit (r = 0.507) – was significantly positive.

H_{11}: There was no relationship between the interpersonal and output process.

The analysis of the correlation matrix showed that the above null hypothesis assumed was not acceptable, as the correlations obtained between the interpersonal factors and all process of output were significant and affected the organization in a positive manner. It was inferred that the correlation of the interpersonal factors with the parameters – i.e., reliability (r = 0.738), competitiveness (r = 0.610), production time (r = 0.554), quality (r = 0.553), and production capacity (r = 0.528) – was significantly positive.

4.4. Regression Analysis

Multiple linear regression analysis has been used for developing regression weights. Table t-value for 10 degrees of freedom at a 5% level is 1.812, and the t-values higher than this in the following table will give the significant parameters.

4.4.1. Production Capacity

The regression outputs for the dependent production capacity variable are shown in Table 4.16.

The regression model developed was significant, as ANOVA analysis F–test = 13.27, p < 05. The predictors identified from the analysis were product concept, raw material and equipment, production planning and control, and quality control parameters of manufacturing competencies; and the strategic agility and management parameters of strategic success.

4.4.2. Production Time

Following are the regression outputs for the dependent variable production time shown in Table 4.17.

TABLE 4.16

Multiple Linear Regression Analysis of the Production Capacity

(a)

Model 1	R	R Square	Adjusted R Square	Std. Error of the Estimate
		Model Summary		
1	.763	.582	.538	.518

Predictors: (Constant), interpersonal, product concept, strategic agility, planning, management, quality control, administration, production planning, raw material and equipment, product design and development, teamwork

(b)

Model	Sum of Squares	Df	Mean Square	F
		ANOVA		
Regression	39.237	11	3.567	13.274
Residual	28.216	105	0.269	
Total	67.453	116		

Predictors: (Constant), interpersonal, product concept, strategic agility, process planning, management, quality control, administration, production planning, raw material and equipment, product design and development, teamwork

Dependent Variable: Production Capacity

(c)

	UN Standardized Coefficients		Coefficients	
	B	Std. Error	Beta	T
(Constant)	1.186	.270		4.495
Product concept	.021	.037	.096	1.550
Product Design and development	-.010	.037	-.061	.270
Process planning	-.005	.027	-.033	.970
Raw material and equipment	.122	.053	.520	2.279
Production planning	-.043	.044	-.198	1.982
Quality control	.111	.037	.521	3.009
Strategic agility	.095	.031	.405	3.081
Management	.041	.025	-.252	1.629
Teamwork	-.005	.036	-.033	.133
Administration	-.097	.051	-.307	.911
Interpersonal	.014	.028	.087	.521

TABLE 4.17

Multiple Linear Regression Analysis of the Production Time

(a)

		Model Summary		
Model	R	R Square	Adjusted R Square	Std. Error of the Estimate
1	.784	.614	.573	.561

Predictors: (constant), interpersonal, product concept, strategic agility, process planning, management, quality control, administration, production planning, raw material and equipment, product design and development, teamwork

(b)

		ANOVA		
Model	Sum of Squares	Df	Mean Square	F
Regression	52.457	11	4.769	15.176
Residual	32.996	105	.314	
Total	85.453	116		

Predictors: (Constant), interpersonal, product concept, strategic agility, process planning, management, quality control, administration, production planning, raw material and equipment, product design and development, teamwork

Dependent Variable: Production Time

(c)

	UN Standardized Coefficients		Coefficients	
	B	Std. Error	Beta	t
(Constant)	.901	.292		3.088
Product concept	.049	.041	.204	1.220
Product design and development	.075	.040	.409	1.882
Process planning	-.099	.029	-.548	3.427
Raw material and equipment	-.013	.058	-.049	.221
Production planning	-.086	.047	-.352	1.818
Quality control	.130	.040	.541	3.256
Strategic agility	.072	.033	.272	2.153
Management	.031	.028	.168	1.129
Teamwork	.052	.039	.319	1.337
Administration	-.033	.055	-.091	.591
Interpersonal	-.026	.030	-.141	.875

The regression model developed was significant as ANOVA analysis showed F – test = 15.17, p < 05. The predictors identified from the analysis was product concept, product design and development, process planning, production planning and control and quality control parameters of manufacturing competencies; and the strategic agility, management, and teamwork parameters of strategic success.

4.4.3. Lead Time

Table 4.18 shows the regression outputs for the dependent variable lead time.

The regression model developed was significant as ANOVA analysis showed F–test = 83, p < 05. The predictors identified from the analysis was product concept, product design and development, raw material and equipment, process planning and control, and quality control parameters of manufacturing competencies; and the management parameter of strategic success.

4.4.4. Quality

Following are the regression outputs for the dependent variable quality. Table 4.19 shows the regression analysis of quality.

The regression model developed was significant as ANOVA analysis showed F – test = 14.9, p < 05. The predictors identified from the analysis was product concept, production planning and control, and quality control parameters of manufacturing competency and management; and the administration parameter of strategic success.

4.4.5. Reliability

Following are the regression outputs for the dependent variable reliability as shown in Table 4.20.

The regression model developed was significant as ANOVA analysis showed F – test = 28.80, p < 05. The predictors identified from the analysis was product concept, product design and development, raw material and equipment, production planning and control, and quality control parameters of manufacturing competencies; and the management and interpersonal parameters of strategic success.

4.4.6. Productivity

Table 4.21 shows the regression outputs for the dependent variable productivity.

The regression model developed was significant as ANOVA analysis showed F – test = 8.90, p < 05. The predictors identified from the analysis

TABLE 4.18

Multiple Linear Regression Analysis of Lead Time

(a)

		Model Summary		
Model	R	R Square	Adjusted R Square	Std. Error of the Estimate
1	.676	.457	.400	.614

Predictors: (Constant), interpersonal, product concept, strategic agility, process planning, management, quality control, administration, production planning, raw material and equipment, product design and development, teamwork

(b)

		ANOVA		
Model	Sum of Squares	df	Mean Square	F
Regression	33.328	11	3.030	8.035
Residual	39.595	105	.377	
Total	72.923	116		

Predictors: (Constant), interpersonal, product concept, strategic agility, process planning, management, quality control, administration, production planning, raw material and equipment, product design and development, teamwork

Dependent Variable: Lead Time

(c)

	UN Standardized Coefficients		Standardized Coefficients	
	B	Std. Error	Beta	T
(Constant)	.714	.320		2.234
Product concept	.017	.044	.077	2.389
Product design and development	.112	.043	.662	2.568
Process planning	-.080	.032	.478	.521
Raw material and equipment	-.074	.063	-.305	1.172
Production planning	-.045	.052	-.200	1.871
Quality control	.031	.044	.139	1.708
Strategic agility	.028	.037	.114	.760
Management	.108	.030	.632	3.578
Teamwork	-.026	.043	-.169	.601
Administration	.043	.060	.129	.706
Interpersonal	-.003	.033	-.019	.097

TABLE 4.19

Multiple Linear Regression Analysis of Quality

(a)

		Model Summary		
Model	R	R Square	Adjusted R Square	Std. Error of the Estimate
1	.782	.611	.570	.482

Predictors: (Constant), interpersonal, product concept, strategic agility, process planning, management, quality control, administration, production planning, raw material and equipment, product design and development, teamwork

(b)

		ANOVA		
Model	Sum of Squares	Df	Mean Square	F
Regression	38.237	11	3.476	
Residual	24.245	105	.232	14.993
Total	62.581	116		

Predictors: (Constant), interpersonal, product concept, strategic agility, process planning, management, quality control, administration, production planning, raw material and equipment, product design and development, teamwork

Dependent Variable: Quality

(c)

	UN Standardized Coefficients		Standardized Coefficients	
	B	Std. Error	Beta	T
(Constant)	1.031	.251		4.123
Product concept	.033	.035	.161	1.956
Product design and development	.049	.034	.313	.933
Process planning	-.016	.025	-.103	.640
Raw material and equipment	.027	.050	.120	.544
Production planning	-.056	.041	-.270	2.388
Quality control	-.011	.034	-.053	1.819
Strategic agility	.123	.029	.542	.274
Management	-.081	.024	-.510	3.415
Teamwork	.039	.033	.278	.169
Administration	.057	.047	.187	1.207
Interpersonal	.015	.026	.095	.587

TABLE 4.20

Multiple Linear Regression Analysis of Reliability

(a)

		Model Summary		
Model	R	R Square	Adjusted R Square	Std. Error of the Estimate
1	.867	.751	.725	.472

Predictors: (Constant), interpersonal, product concept, strategic agility, process planning, management, quality control, administration, production planning, raw material and equipment, product design and development, teamwork

(b)

		ANOVA		
Model	Sum of Squares	df	Mean Square	F
Regression	70.624	11	6.420	28.839
Residual	23.376	105	.223	
Total	94.000	116		

Predictors: (Constant), interpersonal, product concept, strategic agility, process planning, management, quality control, administration, production planning, raw material and equipment, product design and development, teamwork

Dependent Variable: Reliability

(c)

	UN Standardized Coefficients		Standardized Coefficients	
	B	Std. Error	Beta	t
(Constant)	.238	.246		.970
Product concept	.074	.034	.294	2.184
Product design and development	-.050	.033	-.261	.997
Process planning	.081	.024	.426	1.317
Raw material and equipment	060	.049	.217	1.230
Production planning	-.101	.040	-.391	2.521
Quality control	-.031	.034	-.122	1.918
Strategic agility	.003	.028	.010	.095
Management	-.032	.023	-.164	1.373
Teamwork	.096	.033	.558	.930
Administration	-.091	.046	-.242	.956
Interpersonal	.113	.025	.576	2.558

TABLE 4.21

Multiple Linear Regression Analysis of Productivity

(a)

		Model Summary		
Model	R	R Square	Adjusted R Square	Std. Error of the Estimate
1	.695	.483	.429	.708

Predictors: (Constant), interpersonal, product concept, strategic agility, process planning, management, quality control, administration, production planning, raw material and equipment, product design and development, teamwork

(b)

		ANOVA		
Model	Sum of Squares	Df	Mean Square	F
Regression	49.149	11	4.468	8.908
Residual	52.663	105	.502	
Total	101.812	116		

Predictors: (Constant), interpersonal, product concept, strategic agility, process planning, management, quality control, administration, production planning, raw material and equipment, product design and development, teamwork

Dependent Variable: Productivity

(c)

	UN Standardized Coefficients		Standardized Coefficients	
	B	Std. Error	Beta	t
(Constant)	.555	.369		1.506
Product concept	-.016	.051	-.059	2.303
Product design and development	.048	.050	.242	.963
Process planning	-.043	.037	-.220	1.988
Raw material and equipment	-.042	.073	-.147	.580
Production planning	-.114	.060	-.428	1.909
Quality control	.115	.050	.439	2.282
Strategic agility	.089	.042	.308	.105
Management	.015	.035	.076	2.442
Teamwork	.102	.049	.569	.071
Administration	.117	.070	.300	1.681
Interpersonal	-.093	.038	-.456	2.449

were product concept, process planning, production planning and control, and quality control parameters of manufacturing competencies; and the management, administration, and interpersonal parameters of strategic success.

4.4.7. Growth and Expansion

The following Table 4.22 shows the regression outputs for the dependent variable growth and expansion.

The regression model developed was significant as ANOVA analysis showed $F - test = 7.89$, $p < 05$. The predictors identified from the analysis was product concept, product design and development, production planning and control and quality control parameters of manufacturing competency; and the management and administration parameter of strategic success.

4.4.8. Competitiveness

Following are the regression outputs for the dependent variable competitiveness shown in Table 4.23.

The regression model developed was significant as ANOVA analysis showed $F - test = 145$, $p < 05$. The predictors identified from the analysis were product concept, product design and development, production planning and control and quality control parameters of manufacturing competency; and the management and administration parameters of strategic success.

4.4.9. Sales

The regression outputs for the dependent variable sales are depicted in Table 4.24

The regression model developed was significant, as ANOVA analysis showed $F - test = 5.79$, $p < 05$. The predictors identified from the analysis were product concept, production planning and control and quality control parameters of manufacturing competency; and the management and interpersonal parameters of strategic success.

4.4.10. Profit

The following regression analysis for the dependent variable profit is given in Table 4.25.

The regression model developed was significant, as ANOVA analysis showed $F - test = 10.22$, $p < 05$. The predictors identified from the analysis were product concept, raw material and equipment, production planning and control and quality control parameters of the manufacturing

TABLE 4.22

Multiple Linear Regression Analysis of Growth and Expansion

(a)

Model Summary				
Model	R	R Square	Adjusted R Square	Std. Error of the Estimate
1	.673	.453	.395	.662

Predictors: (Constant), interpersonal, product concept, strategic agility, process planning, management, quality control, administration, production planning, raw material and equipment, product design and development, teamwork

(b)

ANOVA				
Model	Sum of Squares	Df	Mean Square	F
Regression	38.011	11	3.456	7.890
Residual	45.989	105	.438	
Total	84.000	116		

Predictors: (Constant), interpersonal, product concept, strategic agility, process planning, management, quality control, administration, production planning, raw material and equipment, product design and development, teamwork

Dependent Variable: Growth & Expansion

(c)

	UN Standardized Coefficients		Standardized Coefficients	
	B	Std. Error	Beta	t
(Constant)	.278	.344		.808
Product concept	.039	.048	.165	1.826
Product design and development	.050	.047	.277	1.071
Process planning	-.064	.034	-.355	.863
Raw material and equipment	-.051	.068	-.194	.744
Production planning	.098	.056	.405	1.858
Quality control	.055	.047	.230	1.160
Strategic agility	.039	.039	.150	.998
Management	.098	.033	.532	3.002
Teamwork	-.022	.046	-.135	.477
Administration	-.069	.065	-.194	1.056
Interpersonal	-.042	.035	-.225	.172

TABLE 4.23

Multiple Linear Regression Analysis of Competitiveness

(a)

		Model Summary		
Model	R	R Square	Adjusted R Square	Std. Error of the Estimate
1	.772	.596	.553	.523

Predictors: (Constant), interpersonal, product concept, strategic agility, process planning, management, quality control, administration, production planning, raw material and equipment, product design and development, teamwork

(b)

		ANOVA		
Model	Sum of Squares	df	Mean Square	F
Regression	42.284	11	3.844	14.59
Residual	28.708	105	.273	
Total	70.991	116		

Predictors: (Constant), interpersonal, product concept, strategic agility, process planning, management, quality control, administration, production planning, raw material and equipment, product design and development, teamwork

Dependent Variable: Competitiveness

(c)

	UN Standardized Coefficients		Standardized Coefficients	
	B	Std. Error	Beta	T
(Constant)	.182	.272		.670
Product concept	.062	.038	.282	1.944
Product design and development	-.015	.037	-.088	1.933
Process planning	.030	.027	.180	.099
Raw material and equipment	.040	.054	.166	.741
Production planning	.028	.044	.125	2.388
Quality control	-.112	.037	-.510	1.819
Strategic agility	-.010	.031	-.042	.325
Management	-.007	.026	-.041	3.415
Teamwork	.048	.036	.320	.317
Administration	.017	.051	.052	1.207
Interpersonal	.057	.028	.336	.038

TABLE 4.24

Multiple Linear Regression Analysis of Sales

(a)

		Model Summary		
Model	R	R Square	Adjusted R Square	Std. Error of the Estimate
1	.615	.378	.312	.721

Predictors: (Constant), interpersonal, product concept, strategic agility, process planning, management, quality control, administration, production planning, raw material and equipment, product design and development, teamwork

(b)

		ANOVA		
Model	Sum of Squares	Df	Mean Square	F
Regression	33.115	11	3.010	5.792
Residual	54.578	105	.520	
Total	87.692	116		

Predictors: (Constant), interpersonal, product concept, strategic agility, process planning, management, quality control, administration, production planning, raw material and equipment, product design and development, teamwork

Dependent Variable: Sales

(c)

	UN Standardized Coefficients		Standardized Coefficients	
	B	Std. Error	Beta	t
(Constant)	.577	.375		1.537
Product concept	-.068	.052	-.279	1.312
Product design and development	.007	.051	.040	.144
Process planning	.011	.037	.061	.302
Raw material and equipment	-.016	.074	-.060	.217
Production planning	-.013	.061	-.054	2.220
Quality control	-.084	.051	-.345	1.934
Strategic agility	.128	.043	.479	.987
Management	.015	.035	.078	2.411
Teamwork	.085	.050	.515	.819
Administration	.113	.071	.312	.992
Interpersonal	-.042	.039	-.222	1.087

TABLE 4.25

Multiple Linear Regression Analysis of Profit

(a)

Model Summary				
Model	R	R Square	Adjusted R Square	Std. Error of the Estimate
1	.719	.517	.466	.715

Predictors: (Constant), interpersonal, product concept, strategic agility, process planning, management, quality control, administration, production planning, raw material and equipment, product design and development, teamwork

(b)

ANOVA				
Model	Sum of Squares	df	Mean Square	F
Regression	57.515	11	5.229	10.221
Residual	53.716	101	.512	
Total	111.231	116		

Predictors: (Constant), interpersonal, product concept, strategic agility, process planning, management, quality control, administration, production planning, raw material and equipment, product design and development, teamwork

Dependent Variable: Profit

(c)

	UN Standardized Coefficients		Standardized Coefficients	
	B	Std. Error	Beta	T
(Constant)	.195	.372		.524
Product concept	-.045	.052	-.165	1.880
Product design and development	-.011	.051	-.053	.219
Process planning	-.006	.037	-.029	.163
Raw material and equipment	-.016	.074	-.054	2.220
Production planning	-.069	.060	-.248	1.144
Quality control	-.105	.051	-.384	2.064
Strategic agility	.095	.043	-.314	.218
Management	.104	.035	.491	2.948
Teamwork	.120	.050	.641	.415
Administration	.101	.070	.249	1.443
Interpersonal	-.042	.038	-.196	1.089

TABLE 4.26

Multiple Linear Regression Analysis of Market Share

(a)

		Model Summary		
Model	R	R Square	Adjusted R Square	Std. Error of the Estimate
1	.721	.520	.469	.477

Predictors: (Constant), interpersonal, product concept, strategic agility, process planning, management, quality control, administration, production planning, raw material and equipment, product design and development, teamwork

(b)

		ANOVA		
Model	Sum of Squares	Df	Mean Square	F
Regression	25.826	11	2.348	10.329
Residual	23.866	105	.227	
Total	49.692	116		

Predictors: (Constant), interpersonal, product concept, strategic agility, process planning, management, quality control, administration, production planning, raw material and equipment, product design and development, teamwork

Dependent Variable: Market Share

(c)

	UN Standardized Coefficients		Standardized Coefficients	
	B	Std. Error	Beta	T
(Constant)	1.025	.248		4.141
Product concept	-.001	.034	-.007	2.037
Product design and development	.026	.034	.185	.761
Process planning	-.099	.025	-.716	4.014
Raw material and equipment	.107	.049	.530	.167
Production planning	.181	.040	.972	4.604
Quality control	-.160	.034	-.874	4.813
Strategic agility	.028	.028	.138	.979
Management	.046	.023	.324	1.950
Teamwork	-.036	.033	-.288	1.089
Administration	-.060	.047	-.219	.272
Interpersonal	.070	.026	.494	.753

TABLE 4.27

Multiple Linear Regression Analysis of Customer Base

(a)

		Model Summary		
Model	R	R Square	Adjusted R Square	Std. Error of the Estimate
1	.601	.361	.294	.657

Predictors: (Constant), interpersonal, product concept, strategic agility, process planning, management, quality control, administration, production planning, raw material and equipment, product design and development, teamwork

(b)

		ANOVA		
Model	Sum of Squares	Df	Mean Square	F
Regression	25.566	11	2.324	5.384
Residual	45.323	105	.432	
Total	70.889	116		

Predictors: (constant), interpersonal, product concept, strategic agility, process planning, management, quality control, administration, production planning, raw material and equipment, product design and development, teamwork

Dependent Variable: Customer Base

(c)

	UN Standardized Coefficients		Standardized Coefficients	
	B	Std. Error	Beta	T
(Constant)	.660	.342		1.930
Product concept	.000	.047	.002	2.007
Product design and development	.030	.047	.183	.656
Process planning	-.061	.034	-.372	.806
Raw material and equipment	-.050	.068	-.207	.734
Production planning	.114	.056	.512	2.056
Quality control	-.047	.047	-.215	1.005
Strategic agility	.049	.039	.202	.940
Management	-.077	.032	.458	2.388
Teamwork	-.066	.046	-.441	.445
Administration	.062	.065	.190	.955
Interpersonal	.045	.035	.265	.279

competencies; and management, administration, and interpersonal parameters of strategic success.

4.4.11. Market Share

Table 4.26 shows the regression analysis of market share. The regression model developed was significant, as ANOVA analysis showed F – test = 10.32, p < 05.

The predictors identified from the analysis was product concept, process planning, production planning and control, and quality control parameters of manufacturing competencies; and management and teamwork parameters of strategic success. The following were the regression outputs for the dependent variable market share.

4.4.12. Customer Base

Table 4.27 shows the regression outputs for the dependent variable customer base.

The regression model developed was significant, as ANOVA analysis showed F – test = 5.38, p < 05. The predictors identified from the analysis were product concept, production planning and control and quality control parameters of manufacturing competencies; and the management parameter of the strategic success.

5

Case Studies in Manufacturing Industries

To validate the data collected through empirical study, four case studies have been conducted in selected industrial units. The main concentration of these case studies has been to search and analyze various aspects of organization function. It involves the problems faced by firms in the era of globalization and need for technology advancement and the role of manufacturing competency in improving their performance. This chapter deals with the case studies conducted at the following automobile manufacturing units:

1. Two-Wheeler Manufacturing Unit
2. Four-Wheeler Manufacturing Unit
3. Heavy Vehicle Manufacturing Unit
4. Agricultural Manufacturing Unit

The data acquired from these industries is of the last five years, that is, 2010–11 to 2014–15. In graphs and charts, the representation is done as: 1st year – 2010–11; 2nd year – 2011–12; 3rd year – 2012–13; 4th year – 2013–14; and 5th year – 2014–15.

5.1. Case Study at the Two-Wheeler Manufacturing Unit

The two-wheeler manufacturing unit is the largest manufacturer of two-wheelers globally. It started its operations in India in 2001 at Manesar (Haryana) and has acquired over 12 million customers in its 12 years of operations. Now, it is recognized as the fastest growing two-wheeler company in India.

The unit is the most important among the automobile sector that has experienced significant changes over the years. It consists of three segments, which are motorcycles, mopeds, and scooters. The key manufacturers in this sector are TVS, Yamaha, Hero, and Bajaj. Its governance structure in shown in Figure 5.1.

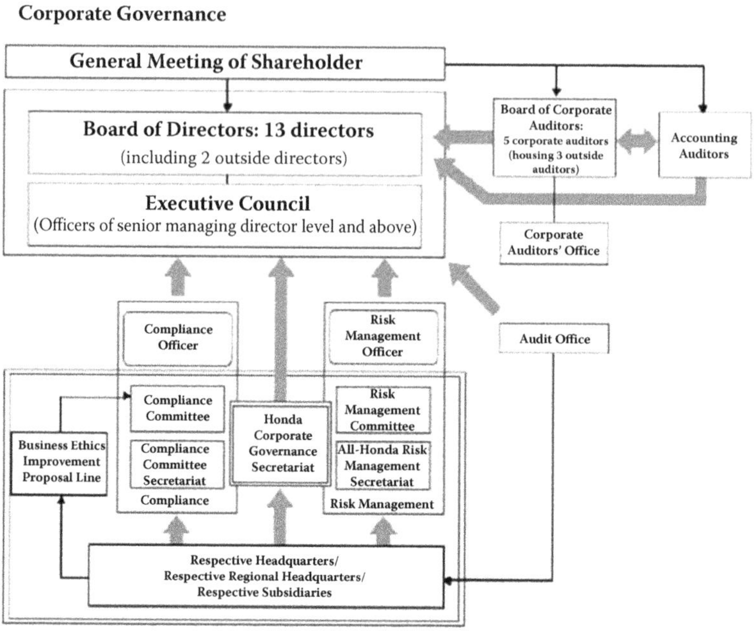

Corporate Governance

FIGURE 5.1
Corporate Governance Structure.

5.1.1. Company Strategy

The principle that this organization follows is implemented by all its companies worldwide.

5.1.1.1. Company Principle (Mission Statement)

Employees are committed to provide high-quality products at reasonable prices in order to satisfy customers worldwide.

5.1.1.2. Fundamental Beliefs

5.1.1.2.1. Respect for the Individual

- **Equality**
  Equality means to respect and recognize differences in one another and provide fair treatment for everyone. Figure 5.2 depicts the unit's philosophy regarding equality.

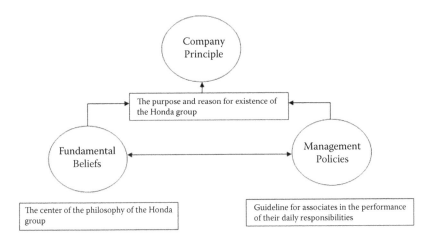

FIGURE 5.2
Two-Wheeler Manufacturing Unit Philosophy.

- **Initiative**
  Initiative implies thinking creatively and acting on one's own judgment without being bound by preconceived notions and the belief that one must be responsible for the results of their actions.
- **Trust**
  Mutual trust is the basis for the relationship between personnel. Trust is created by helping each other, sharing knowledge, recognition, and making an effort in fulfilling one's responsibilities.

5.1.1.2.2. Management Policies
- Always have ambition and youthfulness
- Make effective use of time and develop new ideas
- Enjoy work and encouraging open communication
- Consistently strive for a harmonious work flow
- Be mindful of the value of endeavors and research

Figure 5.3 depicts principle initiatives in product development. Principle initiatives in product development are shown in Figure 5.3.

5.1.1.2.3. Innovation in Manufacturing: Strengthening the Fundamentals
To meet demand, the unit is pursuing innovations in manufacturing technology. They meet the demands and the expectations of customers and their stakeholders.

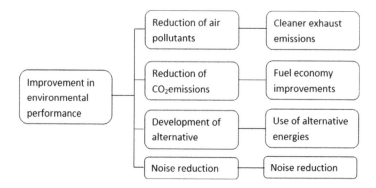

FIGURE 5.3
Principle Initiatives in Product Development.

5.1.1.2.4. The Two-Wheeler Manufacturing Unit's Vision of Environmental Technology

Implementing different technologies, the unit is delivering on the promise of genuine value in environmental responsibility and driving pleasure. Figures 5.4 and 5.5 shows principle initiatives in production and recycling, respectively. Table 5.1 gives 3R for recycling.

5.1.2. Initiatives towards Technological Competency

5.1.2.1. Cutting-Edge Technology

The fundamental design philosophy strives to maximize comfort and space for people, while minimizing the space for mechanical components. With this in mind, its research and development activities include fundamental research and product-specific development.

5.1.2.2. Combi Break System

Generally, it is quite difficult to control a two-wheeler while braking during bad road conditions and emergencies. The combi break system allows easy and simultaneous operation of the rear and front brake while also providing optimal braking performance.

5.1.2.3. Automatic Transmission

The efficient, compact, and oil pressure-controlled transmission is globally the first fully automatic transmission system, delivering a dynamic combination of torque and accelerator response for a superior driving experience.

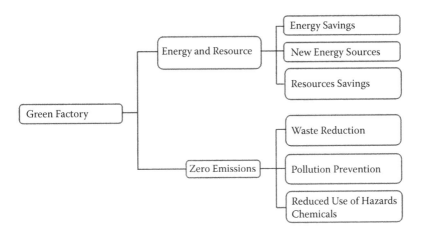

FIGURE 5.4
Principle Initiatives in Production.

FIGURE 5.5
Principle Initiatives in Recycling.

TABLE 5.1

Recycling (3R)

	Development	Products	Use	Disposal
Reduce	Design for reduction			
Reuse	Design for reusability and recyclability	Recycled/reused parts		
Recycling	Recycling and recovery of bumpers			Recycling of IMA batteries
		Recycling of by-products		Compliance with the end-of-life vehicle recycling law in Japan
	Reduction in hazardous or toxic substance			Voluntary measures for recycling motorcycle

5.1.2.4. Fuel Injection System

The fuel injection system is designed to realize ideal combustion, resulting in maximum power output and improved fuel efficiency, while staying environmentally friendly.

5.1.2.5. Idle Stop System

The unit has developed an advanced idle stop system, as shown in Figure 5.6, which reduces fuel consumption and totally blocks out toxic exhaust gases and any unwanted noise. As soon as the vehicle stops, the engine is stopped automatically. When the engine restarts, when the throttle is opened and takes off smoothly.

5.1.3. Management Initiatives

1. *Respecting Independence*: (Challenge)
The unit expects associates to express their independence and individuality. At present, associates are encouraged to think, act, and accept responsibility. Anyone with proposals and ideas should express them. Figure 5.7 shows principles of personnel management.

2. *Ensuring Fairness*: (Equal Opportunity)
The unit offers a simple system with fair rewards for anyone having the same abilities in handling similar sort of work and producing similar results with no concern for nationality or race or gender, making no distinctions on educational basis or career history, and objectively assessing each individual's strengths and aptitudes.

3. *Fostering Mutual Trust*: (Sincerity)
The unit believes that mutual respect and tolerance lays the foundation for trust that binds the employee and the company.

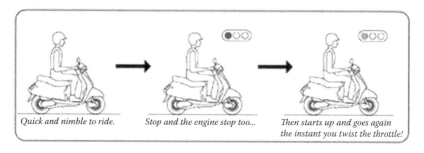

Quick and nimble to ride.   *Stop and the engine stop too...*   *Then starts up and goes again the instant you twist the throttle!*

FIGURE 5.6
Idle Stop System.

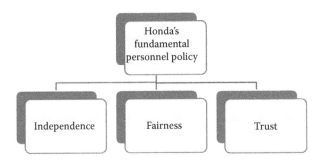

FIGURE 5.7
Principles of Personnel Management.

5.1.3.1. *Fun Expansion*

This is the first industry to promote environmentally-friendly practices in India. Since five years, it has expanded popular initiatives such as Asia Cup, One Make Race, Gymkhana, and Racing Training by Moto GP riders from Japan.

5.1.3.2. *Environment Conservation*

On the environmental front, the unit believes that tomorrow should be better than today. For securing a better environment for the next generation, the unit has implemented several practices. It makes various efforts like reusing and reducing waste for achieving zero emissions; improved efficiency; promotion of green factory, supplier, and dealer initiatives; and resource conservation. Table 5.2 gives the details of the unit's plants, Table 5.3 depicts motorcycle production activities and their targets.

5.1.3.3. *Quality Assurance*

The unit has established quality innovation centers so that quality issues do not arise and enhance the capacity to resolve problems whenever they arise. Specialized departments at these centers are fully equipped to handle cases globally. They provide the resolution of any quality issues, rapid information, and timely diagnosis. They also keep technicians and customers up to date by providing the updates on recommended maintenance techniques. The unit's quality circle is shown in Figure 5.8.

5.1.4. Impact of Competencies on Strategic Success of the Two-Wheeler Manufacturing Unit

Being the world's leading manufacturer in the two-wheeler sector, the unit is working to increase fuel efficiency and lower emissions. People here are

TABLE 5.2

Plant Location

	First Plant	**Second Plant**	**Third Plant**
Location	IMT Manesar, Dist. Gurgaon, Haryana	Tapukara Industrial Area, Dist. Alwar, Rajasthan	Narsapura Area, Dist. Kolar, Karnataka
Annual Production Capacity	16-lac units (at full production)	12-lac units (at full production)	12-lac units (production started) **Additional 6-lac units * (by FY'14 end)**
Lot Size	210,000 m^2	240,000 m^2	96 acres

TABLE 5.3

Motorcycle Production Activities

Exhaust emission (HC)	Fuel economy
Reduce total emissions from new motorcycles by two-thirds	Improve average fuel economy by 30%

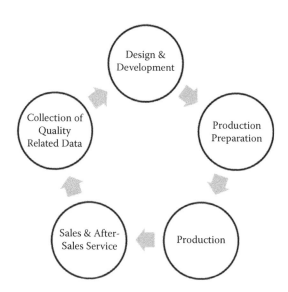

FIGURE 5.8
Two-Wheeler Manufacturing Unit Quality Circle.

working hard to improve environmental performance. For this, programmed fuel injection (PGM-FI) has been implemented. It adapts to changes in engine load caused due to acceleration and deceleration, driving conditions, and adjusting the volume and timing of fuel injection, as well as the timing of ignition for optimal electronic control. With this, fuel efficiency is improved and emissions are reduced without compromising on performance. Table 5.4 depicts the sales plan for last five years.

Production planning deals with ideas of production and execution of production activities. Production control utilizes different types of control techniques for achieving optimum performance out of the production system, thus attaining overall production planning targets. Production planning and control addresses problems of low productivity, inventory management, and resource utilization. From Figure 5.9, it is quite clear that there is an improvement in production activities along with a reduction in equipment breakdown. This is because of better strategizing and planning, thus, production planning and control is an important factor.

Quality control relates to the overall quality of the product produced. From Figure 5.10, one can see the quality of the products have improved continuously and there is reduction in non-confirmatory products, thus leading to an increased number of sales (from Figure 5.14), and an improved profit (from Figure 5.15). This has happened because of better quality provided by the organization.

Product concept is another significant factor. From Figure 5.11, it is evident that the unit's research and development expenditure has increased over the years, which shows that the organization is quite serious towards better ideas so as to improve sales and profit, thus attracting customers. Product design and development is closely related to product concepts such as designing and testing prototypes, which also falls under research and development. Product concept is part of the research and development department of the organization, as product concept involves idea generation either by modifying the existing product or creating a new product. Figure 5.12 shows Capital and Recurring Expenditure.

TABLE 5.4

Sales Plan

		1st year (2010–11)	2nd year (2011–12)	3rd year (2012–13)	4th year (2013–14)	5th year (2014–15)
HMSI Sales	SC	666,450	751,900	907,421	1,242,975	1,465,267
	MC	403,750	520,000	750,632	864,226	980,348
	Total	1,070,200	1,271,900	1,658,053	2,107,201	2,445,615
	Growth	18%	18+19=37%	37+30=67%	67+27=94%	94+16=110%

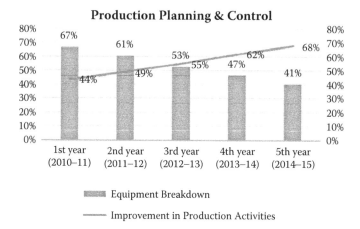

FIGURE 5.9
Production Data for the Last Five Years.

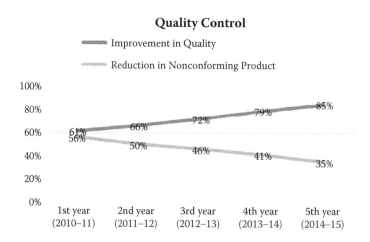

FIGURE 5.10
Quality Control Data for the Last Five Years.

Management support and commitment is another key factor towards the performance of the company. Organization growth has improved over the years, leading to an improved profit. In Figure 5.15, profit has increased over the last five years. From Figure 5.13, the total income of this manufacturing unit is quite clear. Figure 5.16 shows the growth in comparison to previous year.

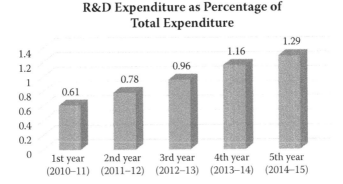

FIGURE 5.11
Research and Development Expenditure as a Percentage of Total Expenditure for the Last
Five Years.

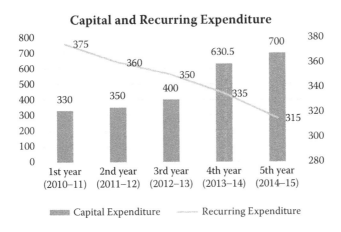

FIGURE 5.12
Capital and Recurring Expenditure for the Last Five Years.

The graph in Figure 5.13 shows the total income in the last five financial years. A gradual rise in total income has been witnessed during this phase. Total income is the sum of the money received, including income from services or employment, payments from pension plans, revenue from sales, or other sources. The sales and profit of the company have been shown in Figures 5.14 and 5.15, respectively, whereas the growth of the company has been represented in Figure 5.16.

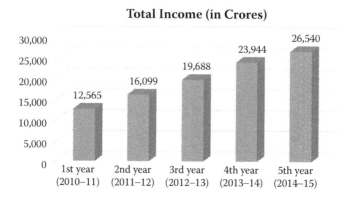

FIGURE 5.13
Total Income for the Last Five Years.

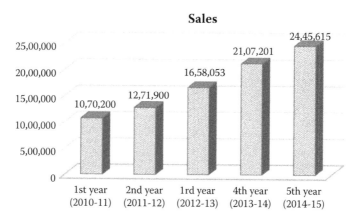

FIGURE 5.14
Sales for the Last Five Years.

Quite similar to the empirical data results, manufacturing competency has an impact on overall performance of the organization, according to case studies. Therefore, it is quite clear that factors obtained from data analysis are same. The factors determined from here are *product concept, production planning and control, quality control,* and *management,* as there has been a considerable improvement in these factors as compared to others.

From the above data, it has been observed that the company sales have been growing since the past five years. This is due to the company introducing new strategies and technologies in their products. It is very

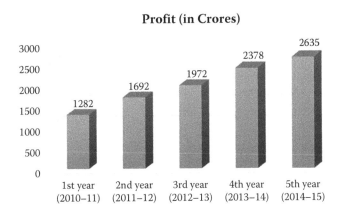

FIGURE 5.15
Profit for the Last Five Years.

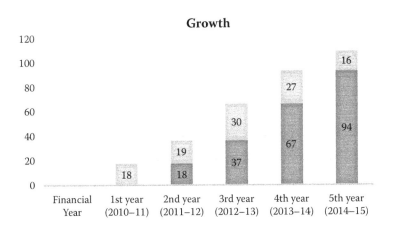

FIGURE 5.16
Growth for the Last Five Years.

difficult to withstand the competitive world of the market otherwise. Every company has to make use of competency in their products at present. If no innovation is made in the product, it becomes obsolete. The unit has a legacy of cutting-edge research and development, resulting in customer-oriented products. Today, due to the new technical center, India is the center of attention worldwide. They are devoted to delivering the best quality products at reasonable prices and at a faster speed through research and development, engineering, purchasing, quality, and designing.

5.1.5. Growth Framework with Customer Needs

1. Premises/Process

- *Voice-of-the-customer at dealerships*: Evaluating customer feedback and bring it on operations.
- *Process efficiency improvement*: Improve work efficiency by elimination of wasteful operations at individual dealerships
- *Single repair programs*: Ensure that customer's most issues are solved in a single repair. Figure 5.17 shows customer satisfaction initiatives.

2. Product

- *Preemptive prevention, expansion prevention, and incorrect delivery prevention*: Boosting product-service quality.

3. People

- *Developing a comprehensive dealership training system*: Strengthening training programs to improve human resources and skill levels.

5.1.6. The Two-Wheeler Manufacturing Unit 's Vision for the Future

1. Future Initiatives

Economic structural change occurs because of awareness about environmental issues globally and the growth of developing countries have a significant effect on their business activities. With the growth of emerging economies, competition in the market has intensified, and online information is exerting a significant influence on performances. In the future, this will require them to provide tailor-made products to every region of the world faster and at a more affordable rate.

FIGURE 5.17
Customer Satisfaction Initiatives.

2. Triple Zero

Zero CO_2 emissions will be guaranteed by using original renewable energy. Also, zero energy risk and zero waste will also be ensured with the collaboration of local communities. Figure 5.18 depicts triple zero and its coexistence with local communities.

5.2. Case Study at the Four-Wheeler Manufacturing Unit

The four-wheeler manufacturing unit, was established in 1981 with the objectives of modernizing the Indian automobile industry, producing fuel-efficient vehicles, and producing indigenous utility cars for the Indian population. The unit is the leader in the car sector, both in terms of revenue earned and volume of vehicles sold. The unit's production of cars commenced in 1983 with 800 vehicles. By 2004, It had produced over 5 million vehicles. The unit's manufacturing facilities are available at two locations that is, Manesar and Gurgaon. These facilities have the capacity to produce over 700,000 units annually.

5.2.1. Principles

The four-wheeler manufacturing unit adopted the norm of same fabric and color uniform for all its employees thereby, giving them an identity. In order to have no time loss in between shifts, employees reported early for shifts. The unit has an open office system and practices kaizen activities, job-rotation, teamwork, quality circles, and on-the-job training.

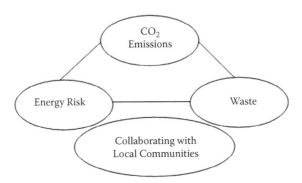

FIGURE 5.18
Triple Zero and Coexistence with Local Communities.

5.2.2. Objectives

There was a need to provide a reliable, better-quality, and cost-effective car for customers. It was established in such a scenario with a resolve to modernize and expand the automobile sector. MSIL was entrusted with the task of achieving the following policy objectives:

- Modernization of Indian automobile industry
- For economic growth, a large volume of vehicles had to be produced
- For conservation of scarce resources, fuel-efficient vehicles were the most pressing need

5.2.3. Company Strategy and Business Initiatives

For three decades, the four-wheeler manufacturing unit has been the world's leader in mini and compact cars. Its technical superiority lies in its capability to pack performance and power into a lightweight and compact engine that is fuel-efficient and clean. It is clearly 'employer of choice' for young managers and automotive engineers across the country.

The unit assures satisfaction among customers. For its earnest efforts, it has been ranked first in this among all Indian car manufacturers for nine consecutive years by J.D. Power Asia Pacific. Figure 5.19 shows the organizational structure of the four-wheeler manufacturing unit.

5.2.4. Technology Initiatives Taken by the Four-Wheeler Manufacturing Unit

Indian customers have a passion for fuel efficiency in choosing automobiles. Achieving more energy per car from a single drop of fuel is a challenge for the designer, but is important for the economy, the planet, and the customer. At the same time, a speed conscious, young, and fast-growing India demands a better pick-up and instant response during acceleration. A third requirement is space efficiency, so that the car can cope with parking lots and congestion on roads.

The unit's new K-series engines deliver on all these fronts. The organization believes that the technology's purpose is to serve mankind with products that use minimum resources to reach out to the maximum number of customers, which is good for their long-term safety, happiness, well-being, and health, while meeting the needs of society. Better technologies, better thinking, better processes, and more ideas helps the unit develop better cars, which in turn helps the customers develop a better quality of life. Today, the research and development team, working shoulder to shoulder with the organizations team, has many achievements:

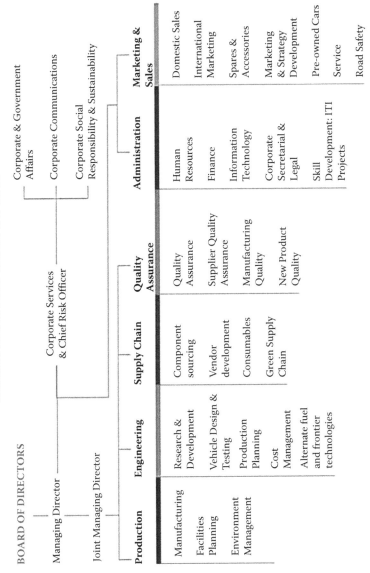

FIGURE 5.19
Organization Structure of the Four-Wheeler Manufacturing Unit.

- The unit launched many new models in India in the last few years
- In India, some of the most fuel-efficient petrol cars come with the organizations badge
- Launch of factory-fitted CNG (compressed natural gas) variants. The factory-fitted CNG vehicles use advanced i-GPI (Intelligent Gas Port Injection) technology. State-of-the-art i-GPI technology is used by the unit's cars.
- The new concept of single minute exchange of dies (SMED) has been adopted. This helps in changing the die setup within a single digit minute, thus, improving operating efficiency and machine utilization.
- Almost all of its cars obey ELV (End of Life Vehicles) norms, which means they can be fully recycled and are free from any hazardous material.
- Plastic intake in the K-series is an example of technologies adopted for lightweight construction. Light piston, nut-less connecting rods, and optimized cylinder block for lightweight configuration, high pressure semi-return fuel system, Smart Distributor Less Ignition (SDLI) with committed advanced injectors and plug top coils for better performance.
- Wagon R Green: Wagon R is a balance of performance, space, and comfort in a new design. Its new model known as Wagon R Green is available on CNG. It ensures fuel-efficiency, safety, reliability, and more power. CNG technology is another step for keeping low cost of ownership for customers.
- ESP (Electronic Stability Program): An onboard microcomputer displays the vehicle's stability and behaviour with sensors on a real-time basis. During instability due to lane change or high-speed cornering, it automatically applies differential brakes at the four wheels to keep the vehicle stable and on the intended track without any additional driver involvement.
- Sequential injection has been introduced in LPG (liquefied petroleum gas) vehicles to ensure reduced emissions, better fuel economy, and improved performance.
- Variable geometry turbocharger (VGT) has been introduced in diesel engines for improving performance and fuel efficiency.

5.2.5. Management Initiatives Taken by the Four-Wheeler Manufacturing Unit

The organization develops a culture in which higher standards of individual's accountability, transparent disclosure, and ethical behavior are ingrained in all business dealings and are shared by the management, employees, and board of directors. The firm has established procedures and systems to ensure that its board of directors are well-equipped and well-informed to fulfil their overall responsibilities and provides the management with strategic direction needed for creating long-term shareholder value.

To meet the organizational responsibilities of safe working environment, the company has established an OHSMS (occupational health and safety management system) for:

- Managing risks: They identify all hazards by undertaking assessments and external and internal audits, as well as all the necessary actions to prevent and control injury, loss, damage, or ill-health.

- Complying with legal and other obligations: They ensure that business here is managed in accordance with occupational health standards and safety legislations.

- Establishing targets and review mechanism: They manage their commitments by using coordinated safety plans and occupational health for each site and area. They tend to measure progress and leadership support, and ensure continual improvement. Health and safety performance is always among parameters for evaluation.

- Providing appropriate training and information: They provide all the necessary tools to all its vendors, employees, visitors, and contractors to ensure safe performance at work.

- Ensuring meaningful and effective consultation: They involve all interested people and employees in the issues that harm health and safety at the workplace.

- Communicating: They believe in the transparent relations of this unit's performance and OHSMS commitments.

- Promoting a culture of safety: They believe that all incidents and injuries can be prevented and everyone is responsible for their own and their processes safety. All responsibilities are clearly defined for all personnel, managers, and supervisors. Figure 5.20 shows the materiality matrix

5.2.6. Quality

The company is awarded ISO 27001 certification by the Standardisation, Testing and Quality Certificate (STQC) from the Indian government. The quality management of is certified against ISO 9001:2008 standard. These systems are reassessed at regular intervals by a third party.

5.2.6.1. Quality Policy

To increase customer satisfaction through improvement of services and products, PDCA functions at the levels of the unit's organization are

			Product safety Business growth & profitability People development & motivation Employee wages & benefits
HIGH		Non-discrimination & human rights Road safety Skill development	Occupational health & safety Customer satisfaction Process emissions Industrial relations Corporate governance Product quality Product emissions Compliance
MEDIUM		Child and forced labour Green supply chain Green service workshop Information security & data privacy Product labeling	Attrition Water conservation Material optimisation Effluent waste Waste management Government policy and regulations Foreign exchange fluctuations
LOW	Biodiversity Indigenous rights	Green products	Competition R & D capability Energy conservation Business ethics Macro economic factors
	LOW	MEDIUM	HIGH

FIGURE 5.20
Materiality Matrix.

followed. Table 5.5 depicts the quality tools employed by MSIL for realizing overall organizational objectives.

5.2.7. Impact of Competencies on Strategic Success of the Four-Wheeler Manufacturing Unit

The unit has recently enhanced its product range with an aim to meet customer needs. A striking new look and a more daring approach to design began with the launch of Swift Dzire and Swift. Another major development is its entry into the used car market, where customers are allowed to bring their vehicles to the 'It True Value' outlet, where it can be exchanged it for a new one by paying the difference.

   Production planning deals with ideas of production and execution of production activities. Production control utilizes different types of control techniques for achieving optimum performance out of the

TABLE 5.5

Quality Tools

5S	4M	3M	3G
SEIRI: Proper selection	**MAN**	**MURI**: Inconvenience	**GENCHI**: Go to actual place
SEITION: Arrangement	**MACHINE** **MATERIAL**	**MUDA**: Wastage	**GENBUTSU**: See the actual thing
SEISO: Cleaning	**METHODS**	**MURA**: Inconsistency	**GENJITS**: Take appropriate action
SEIKETSO: Cleanliness			
SHITSUKE: Discipline			

production system thus attaining overall production planning targets. Production planning and control addresses problems of low productivity, inventory management, and resource utilization. From Figure 5.21, it is quite clear that there is an improvement in production activities along with a reduction in equipment breakdown. This is because of better strategies and planning, thus, production planning and control is an important factor.

Quality control relates to the overall quality of the product produced. From Figure 5.22, it is shown that quality of the products has improved continuously and there is a reduction in non-confirmatory products, thus leading to an increased number of sales (from Figure 5.26) and improved profits (from Figure 5.27). This has happened because of better quality provided by the organization.

Product concept is another significant factor. From Figure 5.23 it is evident that its research and development expenditure has increased over the years which shows that the organization is quite serious towards better ideas so as to improve sales and profit, thus attracting customers. Product design and development is closely related to product concept, as designing and testing prototypes is also done in research and development. Product concept is part of the research and development department of the organization, as product concept involves idea generation either by modifying the existing product or creating a new product.

Management support and commitment is another key factor towards the performance of the company. Organization growth has improved over the years, leading to an improved profit. From Figure 5.25, it is quite clear the improvement in total income. Figure 5.28 shows the growth in comparison to previous year. Figure 5.24 depicts capital and recurring expenditure this unit.

The graph in Figure 5.25 shows the total income by this unit in the last five financial years. A gradual rise in total income has been witnessed during this phase. The total income is the sum of the money received,

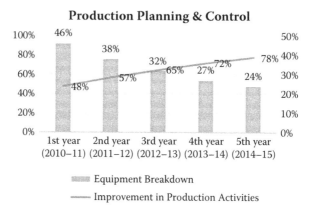

FIGURE 5.21
Production Data for the Last Five Years.

FIGURE 5.22
Quality Control Data for the Last Five Years.

including income from services or employment, payments from pension plans, revenue from sales, or other sources.

The sales and profit of the company have been shown in Figures 5.26 and 5.27, respectively whereas the growth of the company has been represented in Figure 5.28.

Quite similar to the empirical data results, it has been analyzed through the two-wheeler manufacturing unit and four-wheeler manufacturing unit case studies that manufacturing competency has an impact on overall

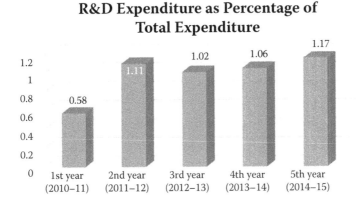

FIGURE 5.23

Research and Development Expenditure as Percentage of Total Expenditure for the Last Five Years.

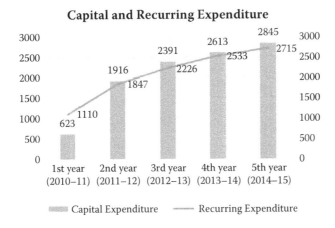

FIGURE 5.24

Capital and Recurring Expenditure for the Last Five Years.

performance of the organization. From the case study, it is quite clear that factors obtained from data analysis are same. The factors determined from here are *product concept, production planning and control, quality control,* and *management* as there has been a considerable improvement in these factors as compared to others.

From the above data, it has been shown that the company sales have been growing. This is due to the introduction of new strategies and technologies in their products, which is a prerequisite for survival in the market.

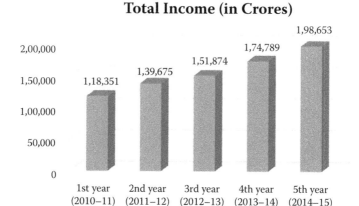

FIGURE 5.25
Total Income for the Last Five Years.

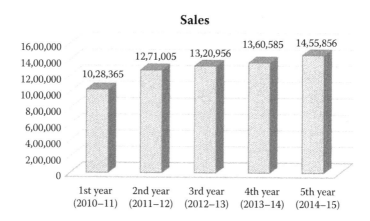

FIGURE 5.26
Sales for the Last Five Years.

5.2.8. Future Plan of Action

- Continuously upgrading existing models
- Developing products with alternative fuel options
- Compliance with safety and emission regulations
- Introducing new technologies
- Developing knowledge of different automotive technologies through standard cost benchmarking and tables.

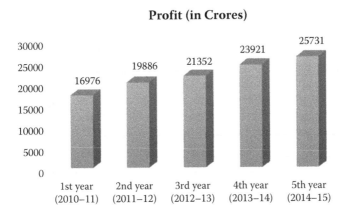

FIGURE 5.27
Profit for the Last Five Years.

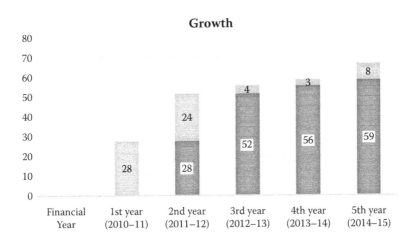

FIGURE 5.28
Growth for the Last Five Years.

5.3. Case Study at the Heavy Vehicle Manufacturing Unit

The heavy vehicle manufacturing unit was established in 1983 for manufacturing LCVs (light commercial vehicles). The project aimed at breaking new ground not only in terms of production technology, but also in building a new value system and culture in the organization, thus enabling it to move forward with confidence into an era of competitive markets.

The commercial vehicle sector in India is currently witnessing significant activity both in meeting challenges in the form of an economic slowdown and on the new product front. The heavy vehicle manufacturing unit is one of the players gearing up for a new destiny. The company's manufacturing plant is located at Ropar in Punjab. Following its pact with commercial vehicle maker of Japan in 2006, the erstwhile is strengthening its presence in the medium and heavy commercial vehicle segment (M and HCV) both in trucks and buses where it was till recently conspicuous by its absence. Future prospects of the M and HCV segment are perceived to be optimistic, though the sector experienced a downfall of 19.13% during the April–December 2012 period.

5.3.1. Technological Initiatives at the Heavy Vehicle Manufacturing Unit

Thus, having recognized the growth potential in Indian market, the unit is putting its shoulder to the wheel at the company. It now has its own director of research and development, who is also a full-time director on the board. The division is currently studying the Indian market and determining potential tailor-made products.

In the last two years, the unit has already invested Rs 14 crores on research and development operations, infrastructure, engine development and testing, and developing equipment. It now plans to add additional equipment and beef up manpower from the current 30 to 100 this fiscal year, as per new project requirements involving new, Japanese designs.

The changed management has, in the meantime, kicked off its entry in the M and HCV market by rolling out two buses and a cargo truck based on the heavy vehicle manufacturing unit chassis.

The unit is also in the process of developing new 16-ton plus cargo trucks for the heavy commercial vehicle segment. The company's existing range spans 5.5–12 tons in cargo and up to 16 tons in buses. The long-term game plan is to become a player in the 5.5–49-ton range, with the heavy vehicle manufacturing unit providing technical inputs for the 12-ton and above category.

5.3.1.1. Clean Diesel Technology

Diesel engines offer numerous advantages like a longer cruising range, low CO_2 emissions, and a superior fuel economy. The unit is focusing on enhancing these advantages and reducing emissions to produce the best diesel engines in the world.

5.3.1.2. Low Pollution Alternative Fuel Vehicles

The unit is devoting itself for developing hybrid-electric trucks and vehicles powered by alternative energy resources, such as dimethyl ether, liquefied petroleum gas, and compressed natural gas. Its low-pollution alternative-fuel vehicles not only contribute to more effective use of limited resources, but also achieve cleaner emissions.

5.3.1.4. *Upgrading the Bread-And-Butter Models*

Current research at Ropar is mostly centred on upgrading branded Mazda products that are called 'bread-and-butter models.' They contribute towards majority of the 13,646 units sold in FY'12, of which passenger carriers accounted for 6,611 units and cargo for 7,035 units. Of these units, this brand chipped in with sales of 154 units. During April–December 2012, sales stood at 8,915 units with passenger carriers pegged at 4,456 units and cargo at 4,459 units. This brand contributed sales of 117 units. The company is expected to record similar numbers in FY'13 due to the on-going slow-down in the CV market, though the company's targeted growth was 10%.

5.3.1.4. *Engine Localization to Trigger Growth*

Meanwhile, the 3.5-liter Mazda engine that is in use since 1984 has under-gone several upgradations at the in-house research and development center to meet BS-IV emission norms with the existing lot of LCVs powered by BS-III and BS-IV engines. Localized Mazda products have enabled the company to notch a 13% market share in the 100,000-unit LCV market it is present in. Several variants of the 3.5-liter Mazda engine have also been developed over the years and the company is now further strengthening its research and development capability for which an invest-ment of INR 200 crores is envisaged.

The research and development division is also undergoing further developments for future products with additional staffers, of which three are Japanese engineers. The CV manufacturer will leverage different series of the 4- and 6-cylinder engines in its M and HCVs that along with transmissions are currently imported from Japan.

With fuel and electronic adjustment carried out to them, these engines will develop different power outputs. For instance, a series of 4-cylinder engines will develop 150–175 horsepower while the 6-cylinder engines will develop 230–300 horsepower and will be fitted in both trucks and buses. These CVs are currently being tested for market to gauge their acceptabil-ity in Indian conditions with imported engines.

The next step will be to forge an agreement for localising the engines. The current unit-badged buses, a 27-seater bus with a 4-cylinder, 5.2-liter engine were followed by the launch of a 45-seater super-luxury bus (LT 134) with a 6-cylinder, 7.7-liter engine. But the company pictures more market potential for the mid-segment 41-seater bus as competitors have done well in this category.

In 2011, the unit had launched a 12.5-ton cargo truck IS12T (with a 4-cylinder, 5.2-liter, 150 hp engine) based on its chassis. While customer response to the cargo truck is good, its price is not acceptable as it is powered by an imported engine. Therefore, the company is selling the vehicle on a nonprofit basis.

With its engine, the buses can compete with Mercedes and Volvo products but cannot compete with the indigenous products made by Tata Motors and Ashok Leyland unless they localize the imported engines and transmissions for its platform products. Branded buses include mini buses, ambulances, school buses, executive buses, and some city buses. Trucks include the crew cab truck, tipper, and cargo carriers, like Sartaj and Cosmo, among others.

CNG power is available on the platform since 2001 and the company claims it has a 70% share in this segment of passenger carriers in Delhi. More variants on the Mazda platform are in the pipeline with a motive of tapping the market in areas where they do not exist at present. The LCV market has also been performing better than other CV segments with the April–December 2012 SIAM results showing a 15% growth in domestic sales for LCVs from 327,406 units during April–December 2011 to 378,509 units in 2012.

In comparison, M and HCV domestic sales dipped 19% to 198,079 units in 2012 from 244,921 units in the earlier year. It is amply clear that it now means business. It has developed a sizeable vendor base of 476 units that will facilitate the future localization process of heavy vehicle manufacturing unit products. Though exports currently form a negligible portion of total production (i.e., 1,000) units per annum head to Bangladesh, Sri Lanka, Nepal, and African countries where the company was earlier exporting tractors, prospects of the product basket expanding in the future are bright.

The unit practices a culture built on the principles of good transparency, corporate governance, and disclosure in all its processes and activities. It gives high priority to ethics and core values. It believes that for a company to be successful, it must consider itself the trustee and custodian of all its stakeholders. It seeks profit and corporate excellence by offering quality services and vehicles to its esteemed customers.

The unit fosters team spirit in employees by continuously raising their participation in decision making. It places high emphasis on lifetime loyalty and integrity to the company. It recognizes that it is focuses on good corporate governance and is rewarded for being better managed enterprise.

5.3.2. Management Initiatives at the Heavy Vehicle Manufacturing Unit

Its mission has been shown in Figure 5.29.

5.3.2.1. Corporate Vision

Pictures itself as a leader in diesel engines, commercial vehicles, and transportation, respecting the environment and supporting their customers.

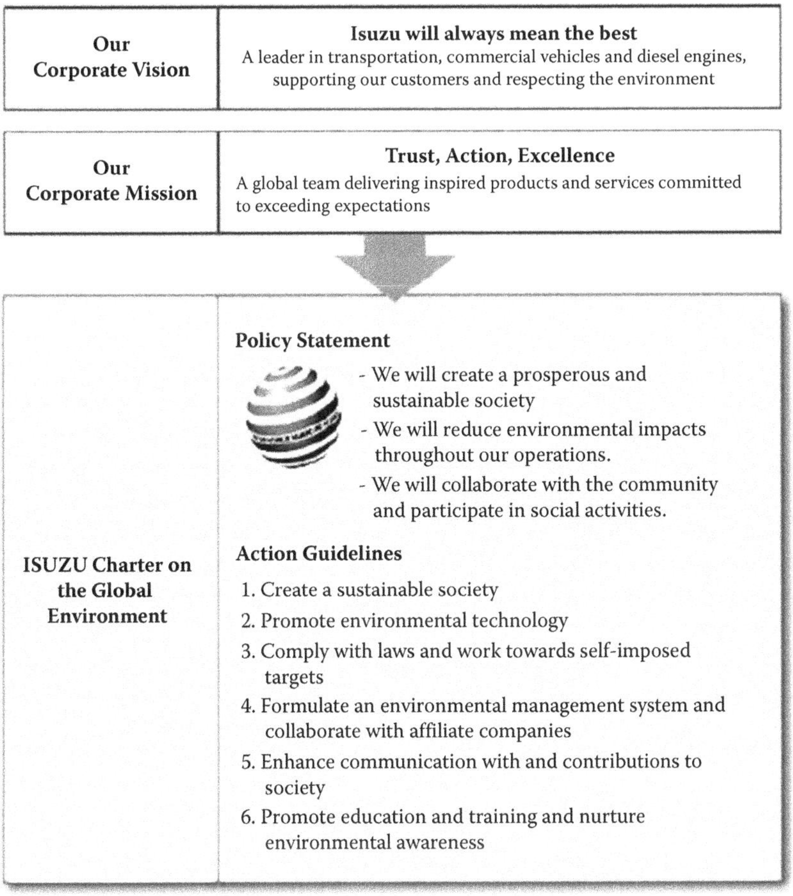

Our Corporate Vision	**Isuzu will always mean the best** A leader in transportation, commercial vehicles and diesel engines, supporting our customers and respecting the environment

Our Corporate Mission	**Trust, Action, Excellence** A global team delivering inspired products and services committed to exceeding expectations

ISUZU Charter on the Global Environment	**Policy Statement** - We will create a prosperous and sustainable society - We will reduce environmental impacts throughout our operations. - We will collaborate with the community and participate in social activities. **Action Guidelines** 1. Create a sustainable society 2. Promote environmental technology 3. Comply with laws and work towards self-imposed targets 4. Formulate an environmental management system and collaborate with affiliate companies 5. Enhance communication with and contributions to society 6. Promote education and training and nurture environmental awareness

FIGURE 5.29
Heavy Vehicle Manufacturing Unit Mission.

5.3.2.2. Corporate Mission

A global team delivering inspired services and products that is committed to exceeding expectations.

5.3.2.3. Focus on Localization

The CV maker has also set up a bus manufacturing plant for luxury and air-conditioned buses. Until recently, the manufacture of buses was out-sourced to external manufacturing plants. It has two major plants in Punjab which develop buses as per the technology, design, and chassis given to them. In 2008, the company had entered the bus market with its designed chassis, on which bus bodies were assembled locally.

Since the earlier bus chassis was based on the Mazda platform, the company operates from two plants – the older one which produces the Mazda chassis for the branded LCVs – while the new plant produces chassis based on their technology for M and HCVs, and bus bodies.

Though initially about 100 units of chassis were imported, they are now being developed in-house leveraging the Japanese design. But the production of old models of buses built by external manufacturers also continues. And for tapping the M and HCV segment in the 12–49 ton range, the focus will be on heavy vehicle manufacturing.

It is also looking at developing low-floor city buses for which it will have to design the relevant chassis. At present, its portfolio spans high-floor and semi-low floor buses. Also, it is not present in bigger city buses like the Marcopolo buses that run on Indian roads. Among the new products in the M and HCV segment that the company has targeted at is a 41-seater, 11-meter front-engine bus (IS12B) that will be positioned between the two buses already launched. Powered by a BS-III 5.2-liter, 4-cylinder engine developing 173 horsepower, it will be equipped with airbags, air suspension and ABS. The CV maker has analyzed the market dynamics and expects that this model, which will be launched in early 2013, has a large market potential.

5.3.3. Quality

5.3.3.1. Quality Policy

The quality policy is shown in Figure 5.30.

5.3.3.2. Winning the Trust of Customers

The unit aimed at winning trust of customers by providing services and products to the society and thus contributing to the creation of a prosperous society.

5.3.3.3. Contributing to Society

The unit has undertaken social contribution activities as a good corporate citizen.

5.3.3.4. Ensuring Harmony with International and Regional Communities

The unit respects the customs and cultures of regions and nations, thereby, contributing to their development through its activities.

FIGURE 5.30
Quality Policy (5S).

5.3.3.5. Making Contribution to Preserving the Environment

The unit works for the preservation and protection of environment not only as corporate citizens residing on earth but also through its business activities, by involving the company with regional and social environmental conservation activities.

5.3.4. Impact of Competencies on Strategic Success of the Heavy Vehicle Manufacturing Unit

The growth strategy is to kick off operations with imported engines and transmissions and slowly localize them over a period of time. But, all this will come at a cost (i.e., a sizeable investment). The company is considering the proposition to localize the 4-liter engine for trucks and buses in the long-term as higher horsepower engines come at an increased cost.

Production planning deals with ideas of production and execution of production activities. Production control utilizes different types of control techniques for achieving optimum performance out of the production system, thus attaining overall production planning targets. Production planning and control addresses problems of low productivity, inventory management, and resource utilization.

From Figure 5.31, it is quite clear that there is an improvement in production activities along with reduction in equipment breakdown. This is because of better strategies and planning, thus, production planning and control is an important factor.

Quality control relates to the overall quality of the product produced. From Figure 5.32, it is shown that the quality of the products has improved continuously and there is a reduction in non-confirmatory products, thus leading to an increased number of sales (from Figure 5.36) and improved profits (from Figure 5.37). This has happened because of better quality provided by the organization.

Product concept is another significant factor. From Figure 5.33, it is evident that the unit's research and development expenditure has increased over the years, which shows that the organization is quite serious towards implementing better ideas to improve sales and profit, thus attracting customers. Product design and development is closely related to product concept, as designing and testing prototypes is also done in research and development, and product concept involves idea

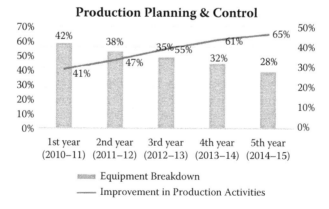

FIGURE 5.31
Production Data for the Last Five Years.

FIGURE 5.32
Quality Control Data for the Last Five Years.

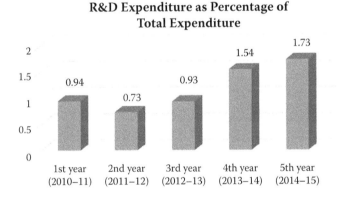

FIGURE 5.33
Research and Development Expenditure as Percentage of the Total Expenditure for the Last Five Years.

generation either by modifying the existing product or creating a new product.

Management support and commitment is another key factor towards the of the company. Organization growth has improved over the years so leading to an improved profit. From Figure 5.35, the improvement in the total income is quite clear. Figure 5.38 shows the growth in comparison to previous year. Figure 5.34 shows capital and recurring expenditure.

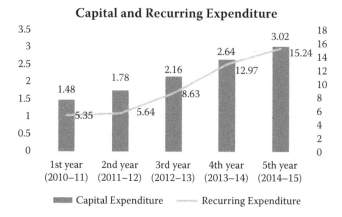

FIGURE 5.34

Capital and Recurring Expenditure for the Last Five Years.

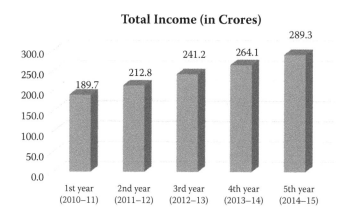

FIGURE 5.35

Total Income for the Last Five Years.

The graph in Figure 5.35 shows the total income in the last five financial years. A gradual rise in total income has been witnessed during this phase. Total income is the sum of the money received, including income from services or employment, payments from pension plans, revenue from sales, or other sources.

The sales and profit of the company have been shown in Figures 5.36 and 5.37, respectively, whereas the growth of the company is represented in Figure 5.38.

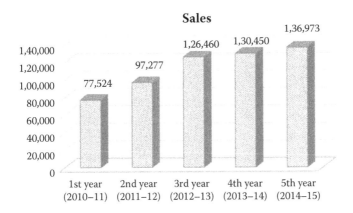

FIGURE 5.36
Sales for the Last Five Years.

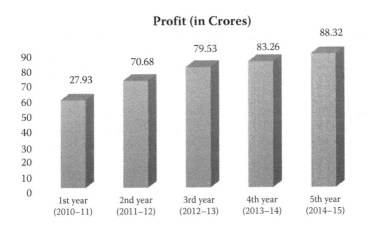

FIGURE 5.37
Profit for the Last Five Years.

Quite similar to the empirical data results in the two-wheeler manufacturing unit, four-wheeler manufacturing unit, and the heavy vehicle manufacturing unit case studies, it is shown that manufacturing competency has an impact on overall performance of the organization. Therefore, factors obtained from data analysis are same. The factors determined from here are *product concept, production planning and control, quality control,* and *management* as there has been a considerable improvement in these factors as compared to others.

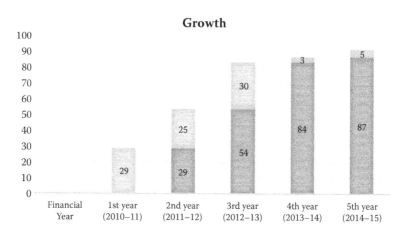

FIGURE 5.38
Growth for the Last Five Years.

5.3.5. Future Initiatives

5.3.5.1. Safety, Economy, and Environment (SEE) Technology for Future Generation

Earning the trust of customers globally is at the heart of development philosophy. The basis for all product development initiatives is SEE technology. The aim is to build advanced technologies with world-class performance in each area. Their unwavering objective is to supply the global market with products that combine economy and safety with a reduced environmental impact. Figure 5.39 shows SEE technology at this unit.

5.4. Case Study at the Agricultural Manufacturing Unit

In 1960s, the prevalence of the green revolution triggered the large-scale use of tractors. The country witnessed an urgent need to build adequate indigenous capacity, in order to meet the increasing demand of tractors.

In 1965, the design and development of tractors based on wisdom and knowledge was initiated by Central Mechanical Engineering Research Institute (CMERI) in Durgapur. It was decided that the name of the product ought to signify India, as well as power and grace, besides being easy to pronounce. In 1970, the Punjab government established this unit.

ISUZU
In pursuit of customers' trust

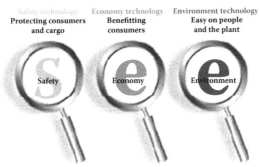

Safety technology Economy technology Environment technology
Protecting consumers Benefitting Easy on people
and cargo consumers and the plant

Safety Economy Environment

Technology

FIGURE 5.39
SEE Technology.

5.4.1. Mission Statement

This company endeavors to create India's largest network for distribution of automobiles, automobile-related products, and services.

5.4.2. Vision Statement

The founders of this company passionately believed that Indians are second to none. People at Mahindra try to prove them correct by believing in themselves and making this unit known globally for the quality of its products and services.

5.4.3. Management Initiatives

The Mahindra group has a core values that are fundamental to all the group companies. At the agricultural manufacturing unit, its core values are influenced by the past, tempered by the present and are designed to shape one's future. Core values are the compass that guides the actions, both corporate and personal.

5.4.3.1. Good Corporate

The agricultural manufacturing unit continues to seek long term success, aligned with the country's needs. They will do this without any compromise on ethical business standards. Figure 5.40 represents core values of this unit.

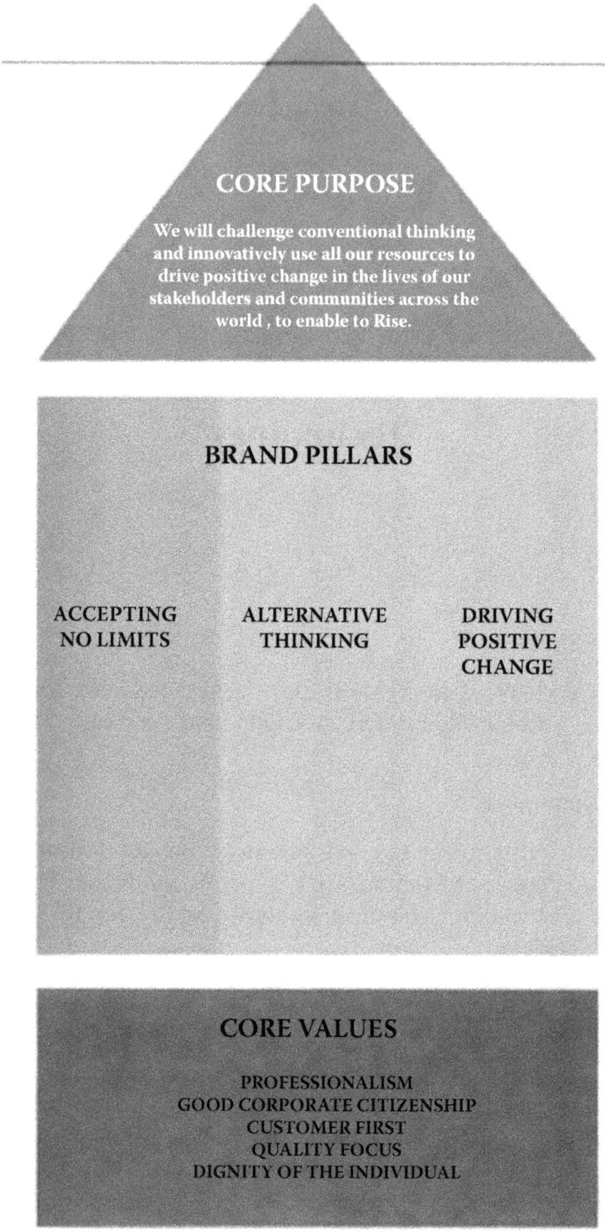

CORE PURPOSE

We will challenge conventional thinking and innovatively use all our resources to drive positive change in the lives of our stakeholders and communities across the world , to enable to Rise.

BRAND PILLARS

ACCEPTING   ALTERNATIVE   DRIVING
NO LIMITS   THINKING   POSITIVE
                                       CHANGE

CORE VALUES

PROFESSIONALISM
GOOD CORPORATE CITIZENSHIP
CUSTOMER FIRST
QUALITY FOCUS
DIGNITY OF THE INDIVIDUAL

FIGURE 5.40
Core Values.

5.4.3.2. Citizenship Professionalism

They sought the best people and give them freedom and opportunity to grow. They support well-reasoned risk-taking and innovation, but demand performance.

5.4.3.3. Customer First

The unit exists and prospers only because of their customers. They respond to their expectations and needs speedily, courteously and effectively. An integrated development strategy at the agricultural manufacturing unit is shown in Figure 5.41.

FIGURE 5.41
Integrated Development Strategy.

5.4.3.4. Quality Focus

Quality is the basis for delivering value for money to the customers. The unit makes quality a driving force in their products, work and while interacting with others.

5.4.3.5. Individual's Dignity

The unit values an individual's dignity, respect the time and efforts of others, and sustains the right to express disparity. Through their actions, they nurture trust, transparency, and fairness.

5.4.3.6. Continuously Improving Systems and Processes

They promote the plan-do-check-act (PDCA) method for analysis and improvement. Emphasis is laid on education and training so that everybody can do their jobs better.

5.4.3.7. Improving Productivity, Safety, Effectiveness, and Reducing Waste by Using Kaizen

The company trains for consistency in reducing variation, building a foundation for common knowledge and allows workers for easy understand their roles.

5.4.3.8. Encouraging Staff to Learn from One Another and Offer an Environment and Culture for Effective Teamwork

The unit makes consistent efforts to implement leadership. It expects managers and supervisors to understand their processes and workers. It instructs to not simply supervise, but to provide resources and support so that each staff member can perform his or her best.

5.4.3.9. Emphasis the Importance of Transformational Leadership and Participative Management

The unit encourages employees not just focus on meeting quotas and targets, but to reach their full potential. It also emphases on eliminating fear.

5.4.3.10. Allowing People to Perform at Their Best by Ensuring, They are Not Afraid to Express Concerns or Ideas

There is an emphasis on doing the right thing versus blaming others when mistakes occur. Workers must be encouraged to find better ways of doing things.

5.4.3.11. Ensuring that Leaders Work with Teams and Act in the Company's Best Interests

Honest and open communication should be employed to allay fear and break down barriers among departments.

5.4.3.12. Building the 'Internal Customer' Concept – Recognize that Each Department Serves Other Departments that Use Their Output

The goal is to build a shared vision and use cross-functional teamwork for creating understanding and reduce adverse relationships.

5.4.3.13. Rather than Measuring the People Behind a Process, Measure the Process

Allow everybody to take pride in their work without any comparison. Treat everyone in same manner, and don't compare with others. Over time, the system will gradually raise the level of everybody's work to a higher level.

5.4.3.14. Enable Self-Improvement by Implementing Education

The target is improving the skills of workers and encouraging them to learn new skills, thus preparing them for future challenges. The workforce becomes more adaptable to changes and achieves improvements.

5.4.4. Technological Initiatives

5.4.4.1. Principles

1. Consistency of purpose to plan services and products, that will have a market and keep the organisation in competition and provide jobs
2. Emphasize short-term profits: short-term thinking and a push from bankers and owners for dividends
3. Mobility of management: job hopping
4. Excessive medical costs
5. Excessive costs of liability.

During the years under study, the farm division while meeting the upcoming engine emission norms, focused on retaining fuel efficiency. Table 5.6 shows the technologies imported by this agricultural manufacturing unit

TABLE 5.6

Technologies Imported by This Unit

S. No	Technology Imported	Year of Import	Status
1	New generation engine management system	2009	Technology absorbed
2	Electronic programs for safety, stability, and steering control	2009	Technology absorbed
3	Controller Area Network (CAN)-based networking	2009	Technology absorbed
4	Advanced material technology	2009	Technology absorbed
5	Development of components using alternate material and advanced manufacturing processes	2010	In process of absorption
6	Engine upgrades and emission improvement technologies	2010	In process of absorption
7	Technology for Noise Vibration Harshness (NVH) Management	2010	Technology absorbed
8	Electrical and electronic technology for safety, infotainment, and convenience feature addition	2010	Technology absorbed
9	New suspension system for improved comfort	2010	Technology absorbed
10	Agri implements technology transfer	2010	In process of absorption
11	Advanced engine technologies	2011	In process of absorption
12	Advanced propulsion technologies	2011	In process of absorption
13	Technology for NVH improvement	2012	In process of absorption
14	Hybrid vehicle technology	2012	In process of absorption

This was done on the engines with improvement focused on technology and overall tractor optimization. Efforts were focused on developing a range of mechanization solutions:

- Global vision with specific focus on exports backed by long-term strategy
- Capability displayed in developing new products
- Low costs for reengineering efforts and improved productivity
- Skilled and motivated workforce
- Capability to develop service and sales network

- Rapidly increasing exports: world-class management and quality systems certified first through ISO 9000

5.4.5. Quality

The main pillars of total quality management are quality assurance systems in manufacturing, new product development, customer operations, sales, and supplier management. On the top of the model stands the purpose of satisfying all employees, suppliers, customers, stakeholders, and society. The involvement of all three major stakeholders through working together improved tremendously.

5.4.6. PDCA Cycle

One of the basic needs of TQM is continuous improvement. In today's environment, if one does not improve, the competitor will take away the market share. This has been the case with European electrical and electronic companies and American automobile manufacturers, who have seen Japanese competitors eating into their market share. Process management and employee suggestions help identify areas for improvement to continuously improve processes, products, and services. An ideal tool is the well-known PDCA cycle. This is also called the 'Deming cycle' as it was developed and promoted by the American quality guru Edward Deming.

The essential elements of PDCA cycle are:

1. Select the theme or project
2. Grasp the present status
3. Analyze the cause and determine corrective action
4. Implement corrective action
5. Take appropriate action
6. Conclusion and future plans
7. Check the effects.

5.4.7. Learning and Development

Training is provided to employees so that they are exposed to new ideas, concepts and expand their horizons. A special emphasis is laid on growth and development of their members. They conduct and participate in training programs and workshops that comprise the managerial, behavioral, and technological growth of their members. Training and development strategy at Swaraj is linked to the strategic plan of the firm. The business plan forms the basis for training and development plans which defines the competence, knowledge, and skills required to meet the organizational objectives.

5.4.8. Impact of Competencies on Strategic Success of the Agricultural Manufacturing Unit

The strategy is to begin operating with imported technologies and slowly localize the technologies over a period of time. Further, considerable research and development work still must be undertaken for making inroads in the tractor segment.

Production planning deals with ideas of production and execution of production activities. Production control utilizes different types of control techniques for achieving optimum performance out of the production system thus attaining overall production planning targets. From Figure 5.42, it is quite clear that there is an improvement in production activities along with reduction in equipment breakdown. This is because of better strategies and planning, thus, production planning and control is an important factor.

Quality control relates to the overall quality of the product produced. From Figure 5.43, it is seen that quality of the products has improved continuously and there is a reduction in non-confirmatory products, thus leading to an increased number of sales (from Figure 5.46) and improved profit (from Figure 5.47). This has happened because of better quality products provided by the organization.

Product concept is another significant factor. From Figure 5.44, it is evident that its development expenditure has increased over the years, which shows that the organization is quite serious towards better ideas so

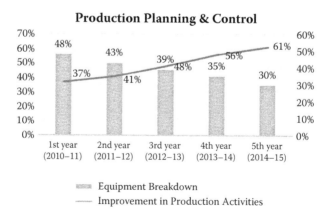

FIGURE 5.42
Production Data for the Last Five Years.

FIGURE 5.43
Quality Control Data for the Last Five Years.

as to improve sales and profit, thus attracting customers. Product design and development is closely related to product concept, as designing and testing prototypes is also done in development. Product concept is part of the development department of the organization, as product concept involves idea generation either by modifying existing products or creating new ones. Efforts were focused on developing a range of mechanization solutions:

- Global vision with specific focus on exports backed by long-term strategy
- Capability displayed in developing new products
- Low costs for reengineering efforts and improved productivity
- Skilled and motivated workforce
- Capability to develop service and sales network
- Rapidly increasing exports: world-class management and quality systems certified first through ISO 9000

Management support and commitment is another key factor towards the performance of the company. Organization growth has improved over the years so leading to an improved profit. From Figure 5.45, it is quite clear the improvement in total income. Figure 5.48 shows the growth in comparison to previous year. Revenue expenditure refers to expenditure concerned with the costs of doing business on a day-to-day basis. When companies make revenue expenditure, the expense offers immediate benefits, rather than long term ones. This is differentiated with capital expenditures, which are long-term investments to help a business grow and thrive.

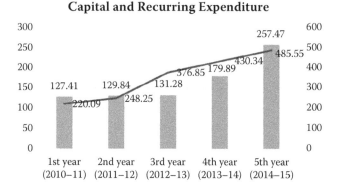

FIGURE 5.44
Expenditure.

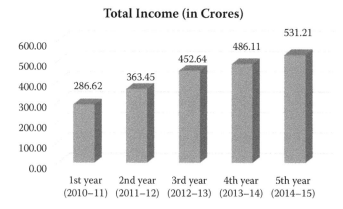

FIGURE 5.45
Total Income for the Last Five Years.

The graph in Figure 5.45 shows the total income in the last five financial years. A gradual rise in total income has been witnessed during this phase. Total income is the sum of the money received, including income from services or employment, payments from pension plans, revenue from sales, or other sources. The sales and profit of the company have been shown in Figures 5.46 and 5.47, respectively, whereas the growth of the company has been represented in Figure 5.48.

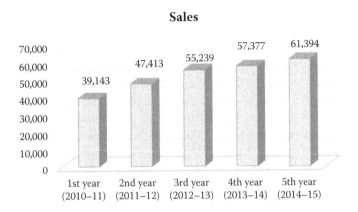

FIGURE 5.46
Sales for the Last Five Years.

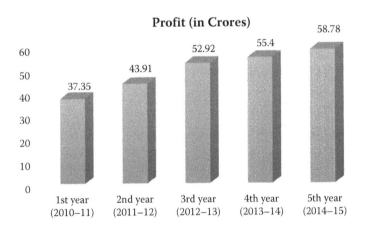

FIGURE 5.47
Profit for the Last Five Years.

Quite similar to the empirical data results that have been analyzed through the two-wheeler manufacturing unit, four-wheeler manufacturing unit, heavy vehicle manufacturing unit, and agricultural manufacturing unit case studies that manufacturing competency has an impact on overall performance of the organization. So, from the case study it is quite clear that factors obtained from data analysis are the same. The factors determined from here are *product concept, production planning and control, quality control*, and *management*, as there has been a considerable improvement in these factors as

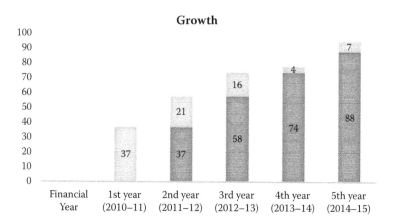

FIGURE 5.48
Growth for the Last Five Years.

compared to others. The above data shows that the performance of the company has been improving. This is due to the introduction of new strategies and technologies in their products. It is mandatory for companies to employ competency lest their products become obsolete.

5.4.9. Future Plan of Action

The unit continues its efforts for development of new technologies and products to meet the ever-growing competitive pressures, regulatory requirements, customer needs, and preparation for the future. It continues to aggressively pursue technology development. Few areas in this direction are weight reduction by using alternative materials, designing modularity, value analysis value engineering (VAVE) approach for meeting cost pressures. Development and adoption of safety technologies also remain a key area.

6

Multi-Criteria Decision-Making Techniques

6.1. Analytical Hierarchy Process (AHP)

AHP is a structured decision-making technique that involves organizing and analyzing multiple criteria by structuring them into a hierarchy and assessing their relative importance. It also helps in the comparison of alternatives for every criteria and defining ranks to the alternatives, as stated by Decision Support System (DSS) Resources. AHP captures both objective and subjective evaluation measures and provides a valuable tool for checking the consistency of the options as suggested by the team, thus reducing favoritism in decision making.

AHP allows organizations to minimize distractions within decision making – like planning, lack of participation, and focus – which are quite costly and prevent them from making the right choice. AHP is very helpful in situations when the decision set involves multiple criteria with ratings according to multiple value choice. It splits the problem into small evaluations while keeping their part in the global verdict. AHP is a multi-attribute decision-making (MADM) technique (Saaty, 1980). It is based on matrices and their corresponding eigenvectors for generating approximate values (Saaty, 1980, 1994).

AHP is a prescriptive and a descriptive model of decision making. It is valid for thousands of applications and the results were accepted and used by the organizations in this study (Saaty, 1994). Thus, presently, it is the most widely used multiple-criteria decision-making (MCDM) technique (Singh and Ahuja, 2012). AHP compares criteria or alternatives in a pairwise mode. For this, AHP uses an absolute numbers scale, validated by theoretical and physical experiments.

AHP can be applied to the following situations (De Steigur et al., 2003):

- Choice: Choosing one alternative from a given set of alternatives, when there are multiple criteria involved

- Ranking: Putting alternatives in an order from most to least desirable or vice-versa
- Prioritization: Determining the relative importance of alternatives, as opposed to selecting only one or simply ranking them
- Resource allocation: Allocating resources between set of alternatives
- Benchmarking: Comparing the processes in own organization with those of other top-performing organizations
- Quality management: Concerning multidimensional aspects of quality improvement
- Conflict resolution: Resolving disputes among parties with incompatible goals or positions.

The applications of AHP to complex situations are in the thousands, and produce results in problems involving selection, priority setting, resource allocation, and planning among choices. Other areas included are total quality management, balanced scorecard, quality function deployment, business process reengineering, and forecasting.

6.1.1. Comparison Scale for Pairwise Comparison

Pairwise comparison is an important stage in AHP for determining priority values of attributes and providing a relative rating for alternatives.

A measurement scale provides the relative importance of each factor by providing numerical judgments corresponding to verbal ones. This scale is a discrete one and ranges from 1 to 9, where 9 shows the highest importance of one factor over the other and 1 describes equal importance among two factors, as shown in Table 6.1. (Singh and Ahuja, 2012)

6.1.2. Pairwise Comparison of Attributes

While comparing attributes, the significance of the jth sub-objective is calculated from with ith one. For this, according to number of variables, as 11 in this study, an 11 × 11 matrix was prepared and used following procedure for filling this:

1. In the matrix, the diagonal elements are kept 1.
2. Values in the upper triangular matrix are filled by using the data compiled through responses from various organizations.
3. For the lower triangular matrix, the upper diagonal values are reciprocated as $a_{ji} = 1/a_{ij}$. Thus, the comparison matrix for different attributes is shown in Table 6.2.

TABLE 6.1

Comparison Scale Used

Intensity	Definition	Explanation
1	Equal Importance	Two factors contribute equally to the objective
3	Moderately More Important	Experience and judgment favor one factor over another
5	Strongly More Important	Experience and judgement strongly favour one factor over another
7	Very Strongly More Important	A factor is strongly favored and its dominance is demonstrated in practice
9	Extremely More Important	The evidence of favouring one factor over another is of the highest possible order of affirmation
2, 4, 6, 8: Intermediate value when compromise is needed		

Source: Singh and Ahuja, 2012

TABLE 6.2

Pairwise Comparison Matrix

	PC	PDD	PP	RME	PPC	QC	SA	MGT	ADM	TW	INT
PC	1	4.00	3.00	2.00	0.25	3.00	2.00	0.33	5.00	6.00	2.00
PDD	0.25	1	3.00	2.00	0.25	0.33	2.00	0.33	6.00	3.00	2.00
PP	0.33	0.33	1	5.00	0.25	0.5	3.00	0.33	4.00	2.00	1.00
RME	0.50	0.50	0.20	1	0.25	0.33	1.00	0.25	3.00	2.00	4.00
PPC	4.00	4.00	4.00	4.00	1	3.00	4.00	2.00	2.00	4.00	3.00
QC	0.33	3.00	2.00	3.00	0.33	1	4.00	3.00	2.00	4.00	4.00
SA	0.50	0.50	0.33	1.00	0.25	0.25	1	0.25	2.00	4.00	2.00
MGT	3.00	3.00	3.00	4.00	0.50	0.33	4.00	1	4.00	3.00	4.00
ADM	0.20	0.17	0.25	0.33	0.50	0.5	0.50	0.25	1	3.00	2.00
TW	0.17	0.33	0.50	0.50	0.25	0.25	0.25	0.33	0.33	1	2.00
INT	0.50	0.50	1.00	0.25	0.33	0.25	0.50	0.25	0.50	0.50	1
SUM	10.78	17.33	18.28	23.08	4.16	9.74	22.25	8.32	29.83	32.5	27

6.1.3. Analysis Using AHP

Analysis in AHP is done by squaring the pairwise matrix and then squaring until the eigenvectors are calculated from the same first iteration: Squaring the matrix is shown in Table 6.2. Table 6.3 shows the resultant matrix.

TABLE 6.3

Matrix After First Iteration

	PC	PDD	PP	RME	PPC	QC	SA	MGT	ADM	TW	INT
PC	10.99	25.81	33.30	45.47	9.06	15.33	42.32	17.20	66.80	71.99	63.07
PDD	8.29	10.98	15.46	28.79	6.99	9.17	23.89	7.21	39.72	47.81	40.89
PP	8.53	11.31	10.96	20.71	5.82	8.10	19.64	6.78	36.61	45.46	45.39
RME	6.48	9.476	11.89	10.98	4.60	6.50	11.92	5.04	19.12	26.97	25.27
PPC	23.89	47.48	51.62	68.41	10.98	27.05	63.50	21.53	100.82	101.5	82.00
QC	18.97	25.46	32.73	43.64	8.56	10.95	40.98	12.98	64.63	63.30	61.65
SA	5.64	8.76	10.61	11.56	3.95	5.99	10.99	4.758	18.14	24.91	25.08
MGT	17.15	29.65	35.28	45.81	9.42	19.72	41.07	10.95	77.65	80.32	63.82
ADM	5.36	7.87	9.09	9.65	3.06	4.84	9.57	4.90	11.00	16.62	17.81
TW	4.10	5.67	7.25	8.42	2.01	3.12	7.98	3.01	10.81	10.99	11.73
INT	4.16	6.77	8.16	11.48	1.93	4.32	9.94	3.04	14.57	14.57	10.99

Eigenvector

$$
\begin{pmatrix}
0.54269 \\
0.31702 \\
0.28541 \\
0.19773 \\
0.82910 \\
0.52394 \\
0.18309 \\
0.58146 \\
0.14963 \\
0.10983 \\
0.12785
\end{pmatrix}
$$

Second iteration: squaring the matrix is shown in Table 6.3. Table 6.4 shows the resultant matrix.

Eigenvector

$$
\begin{pmatrix}
0.54765 \\
0.31992 \\
0.28802 \\
0.19954 \\
0.83668 \\
0.52873 \\
0.18477 \\
0.58678 \\
0.15100 \\
0.11083 \\
0.12902
\end{pmatrix}
$$

TABLE 6.4
Matrix After Second Iteration

	PC	PDD	PP	RME	PPC	QC	SA	MGT	ADM	TW	INT
PC	2872.45	4438.69	5340.62	6974.14	1714.72	2798.22	6494.77	2351.15	10288.61	11912.35	10984.37
PDD	1680.65	2632.94	3168.98	4071.42	989.09	1634.29	3808.05	1397.90	5948.68	6848.58	6318.28
PP	1508.00	2375.16	2872.43	3716.26	881.76	1471.69	3455.14	1260.45	5367.07	6120.42	5609.11
RME	1028.55	1623.39	1954.30	2604.66	607.40	1007.11	2412.34	863.62	3818.11	4320.70	3948.24
PPC	4343.57	6701.10	8116.23	10695.58	2641.43	4223.94	9953.24	3624.27	15759.88	18305.18	16941.94
QC	2712.22	4268.77	5146.2	6750.21	1657.92	2704.39	6287.39	2311.34	9927.79	11542.13	10630.92
SA	956.18	1503.35	1812.14	2405.70	564.55	934.523	2229.17	798.33	3523.71	3996.73	3651.96
MGT	3064.23	4736.07	5724.76	7508.78	1839.67	2955.26	7000.54	2571.01	10964.38	12703.14	11797.83
ADM	769.24	1218.97	1471.95	1972.00	461.75	763.16	1823.53	645.84	2925.15	3311.16	3006.41
TW	564.71	891.247	1074.57	1428.22	344.49	561.79	1326.85	479.83	2117.73	2437.29	2232.94
INT	666.47	1034.23	1251.44	1644.45	406.91	649.80	1534.55	561.04	2434.65	2828.97	2620.24

Third iteration: Table 6.5 provides the values for the results obtained after squaring the matrix in Table 6.4.

$$
\text{Eigenvector}
\begin{pmatrix}
0.54766 \\
0.31992 \\
0.28802 \\
0.19954 \\
0.83670 \\
0.52874 \\
0.18477 \\
0.58679 \\
0.15100 \\
0.11083 \\
0.12902
\end{pmatrix}
$$

Comparing the eigenvector matrix after the second and third iterations, the two are almost same, so further squaring and iterations will generate the same solution. Now let us check the consistency ratio (CR).

6.1.4. Normalization of Matrix

After comparing the matrix, the next priority vector is to be calculated by normalizing the eigenvector of the matrix. For normalizing, each entry in each column is divided by the total of entries in that column. Table 6.6 characterizes the normalized matrix

The normalized value (r_{ij}) is calculated as:

$$r_{ij} = a_{ij} / \sum a_{ij} \qquad\qquad \text{Eq. 6.1}$$

Table 6.7 shows the normalized matrix with priority weights. Further, the approximate value of priority weights (W_1, W_2, W_3, ..., W_j) is obtained as:

$$W_j = 1/n \sum a_{ij} \qquad\qquad \text{Eq. 6.2}$$

The CR is described as the comparison between the random index (RI) and consistency index (CI). If the CR is less than 0.1 (that is, 10%) then judgments are considered to be consistent and acceptable. Table 6.8 gives RI values.

Formulation of CR:

TABLE 6.5

Matrix After Third Iteration

	PC	PDD	PP	RME	PPC	QC	SA	MGT	ADM	TW	INT
PC	81352435.95	127296994.5	153615161.5	2019906640.90	48745995.7	79685605.92	187857263.9	68148885.93	296784604.5	341123264.6	313657961.9
PDD	47527203.03	74368274.08	89743117.14	117993152.1	28474291.79	46550595.83	109738494.5	39811945.71	173353374.5	199249324.7	183209503.6
PP	42788446.83	66959040.81	80797895.35	106232652.3	25632889.36	41909208.92	98799570.22	35842297	156069602.6	179372882.1	164928548.1
RME	29640285.38	46384612.92	55973864.47	73605795.61	17757663.19	29033307.65	68453433.43	24830172.52	108145267.7	124283656.5	114268993.5
PPC	124283423.80	194470861.5	234679518.8	308601518.7	74473267.02	121735016.7	287006245.5	104114569.1	453436384	521175039.6	479214880.20
QC	78539412.27	122896272.2	148306025.7	195015339.9	47061839.56	76929740.97	181369729.5	65796551.1	286532847.4	329339086.2	302825255.7
SA	27446498.46	42950976.53	51830392.41	68156005.41	16443440.77	26884338.49	63385373.95	22992168.9	100137918.4	115083235	105810778.5
MGT	87166795.1	136390651.1	164589915.7	216422776.3	52228317.9	85375463.81	201279677.2	73018540.72	317982681.3	365485832.1	336065142.6
ADM	22428232.47	35099693.47	42356195.84	55702378.32	13437782.94	21970229.29	51802527.45	18789499.02	81844050.01	64055660.44	86474405.91
TW	16462205.66	25761079.77	31087290.95	40881455.87	9864227.248	16125472.65	38020054.09	13791516.79	60068162.85	69037038.82	6376048.97
INT	19164349.99	29987204.14	36187446	47586304.54	11483780.65	18771343.85	44256363.66	16054526.54	69919910.27	80365256.07	73395148.11

TABLE 6.6

Normalized Matrix

	PC	PDD	PP	RME	PPC	QC	SA	MGT	ADM	TW	INT
PC	0.09	0.23	0.16	0.09	0.06	0.31	0.09	0.04	0.17	0.18	0.07
PDD	0.02	0.06	0.16	0.09	0.06	0.03	0.09	0.04	0.20	0.09	0.07
PP	0.03	0.02	0.05	0.22	0.06	0.05	0.13	0.04	0.13	0.06	0.04
RME	0.04	0.03	0.01	0.04	0.06	0.03	0.04	0.03	0.10	0.06	0.15
PPC	0.37	0.23	0.21	0.17	0.24	0.31	0.18	0.24	0.07	0.12	0.11
QC	0.03	0.17	0.10	0.13	0.08	0.10	0.18	0.36	0.07	0.12	0.15
SA	0.04	0.03	0.02	0.04	0.06	0.02	0.04	0.03	0.07	0.12	0.07
MGT	0.27	0.17	0.16	0.17	0.12	0.03	0.18	0.12	0.13	0.09	0.14
ADM	0.02	0.01	0.01	0.01	0.12	0.05	0.02	0.03	0.03	0.09	0.07
TW	0.01	0.02	0.03	0.02	0.06	0.03	0.01	0.04	0.01	0.03	0.07
INT	0.04	0.02	0.05	0.01	0.08	0.03	0.02	0.03	0.017	0.02	0.04
SUM	1	1	1	1	1	1	1	1	1	1	1

TABLE 6.7

Normalized Matrix With Priority Weights

	1	2	3	4	5	6	7	8	9	10	11	Priority Weights
1	0.09	0.23	0.16	0.09	0.06	0.31	0.09	0.04	0.17	0.18	0.07	0.13671
2	0.02	0.06	0.16	0.09	0.06	0.03	0.09	0.04	0.20	0.09	0.07	0.083883
3	0.03	0.02	0.05	0.22	0.06	0.05	0.13	0.04	0.13	0.06	0.04	0.076326
4	0.04	0.03	0.01	0.04	0.06	0.03	0.04	0.03	0.10	0.06	0.15	0.055339
5	0.37	0.23	0.22	0.17	0.24	0.31	0.18	0.24	0.07	0.12	0.11	0.205799
6	0.03	0.17	0.11	0.13	0.08	0.10	0.18	0.36	0.07	0.12	0.15	0.136203
7	0.05	0.03	0.02	0.04	0.06	0.02	0.04	0.03	0.07	0.12	0.07	0.051052
8	0.3	0.17	0.16	0.17	0.12	0.03	0.18	0.12	0.13	0.09	0.15	0.147038
9	0.02	0.01	0.01	0.01	0.12	0.05	0.02	0.03	0.03	0.09	0.07	0.043663
10	0.02	0.02	0.03	0.02	0.06	0.03	0.01	0.04	0.01	0.03	0.07	0.030582
11	0.05	0.03	0.05	0.01	0.08	0.03	0.02	0.03	0.01	0.01	0.04	0.033406

TABLE 6.8

Random Index

N	1	2	3	4	5	6	7	8	9	10	11	12	13	14	15
RI	0	0	0.58	0.89	1.12	1.24	1.32	1.41	1.45	1.49	1.52	1.54	1.56	1.58	1.59

$$CR = CI/RI \qquad \text{Eq. 6.3}$$

The consistency test values are shown in Table 6.9.
Number of comparisons = 55
CR = 9.4%
Principal eigenvalue = 11.43

6.1.5. Priorities

The resulting weights for the criteria based on the above pairwise comparisons are given in Table 6.10.

Table 6.10 presents the results of AHP. The previous work has been validated through this tool as the same factors that were analyzed from the questionnaire analysis and case studies with maximum priority. So, the factors selected for validation by using SEM are *production planning and control, product concept, quality control,* and *management.*

TABLE 6.9

Consistency Test Values

Maximum Eigenvalue	CI	RI	CR
11.43	0.143	1.52	0.094

TABLE 6.10

Priorities of Attributes

Attributes	Priority	Rank
Product Concept	13.67%	3
Product Design and Development	8.38%	5
Process Planning	7.63%	6
Raw Material and Equipment	5.53%	7
Production Planning and Control	20.57%	1
Quality Control	13.62%	4
Strategic Agility	5.10%	8
Management	14.70%	2
Administration	3.05%	11
Teamwork	4.36%	9
Interpersonal	3.34%	10

6.2. Technique for Order of Preference by Similarity to Ideal Solution (TOPSIS)

TOPSIS was developed for solving MCDM problems. This method is based on the idea that the chosen alternative should be farthest from the negative ideal solution (NIS) and nearest to the positive ideal solution (PIS). This is a more realistic form of modeling as compared to non-compensatory methods, in which alternative solutions are included or excluded according to hard cut-offs. For instance, NIS maximizes the cost and minimizes the benefit whereas PIS maximizes the benefit and minimizes the cost. It assumes that each criterion is required to be maximized or minimized.

In this method, options are graded according to ideal solution similarity. The option has a higher grade, if the option is closer to an ideal solution. An ideal solution is a solution that is the best from any circumstances. The algorithm calculates perceived positive and negative ideal solutions based on the range of attribute values available for the alternatives. The best solution is the one with the shortest distance to the positive ideal solution and longest distance from the negative ideal solution, where distances are measured in Euclidean terms. It assumes that we have m alternatives (options) and n attributes/criteria and we have the score of each option with respect to each criterion.

NIS is a technique for ranking different alternatives according to closeness to the ideal solution. The procedure is based on an intuitive and simple idea, that is, the optimal ideal solution with the maximum benefit is obtained by selecting the best alternative which is far from the most unsuitable alternative with minimal benefits. The ideal solution should have a rank of 1, while the worst alternative should have a rank approaching 0. This method considers three types of attributes or criteria:

- Qualitative benefit attributes/criteria
- Quantitative benefit attributes
- Cost attributes or criteria

In this method, two artificial alternatives are hypothesized:

Ideal alternative: The alternative which has the best level for all attributes considered.

Negative ideal alternative: The alternative which has the worst attribute values.

Let X_{ij} score of option i with respect to criterion j

We have a matrix $X = (X_{ij})$ m×n matrix.

Let J be the set of benefit attributes or criteria (more is better) and J' be the set of negative attributes or criteria (less is better). Table 6.11 shows the decision matrix for TOPSIS.

TABLE 6.11

Decision matrix for TOPSIS

	PC	PDD	PP	RME	PPC	QC	SA	MGT	ADM	TW	INT
PC	1	4	3	2	0.25	3	2	0.33	5	6	2
PDD	0.25	1	3	2	0.25	0.03	2	0.33	6	3	2
PP	0.03	0.03	1	5	0.25	0.5	3	0.33	4	2	1
RME	0.5	0.5	0.2	1	0.25	0.03	1	0.25	3	2	4
PPC	4	4	4	4	1	3	4	2	2	4	3
QC	0.03	3	2	3	0.03	1	4	3	2	4	4
SA	0.5	0.5	0.33	1	0.25	0.25	1	0.25	2	4	2
MGT	3	3	3	4	0.5	0.33	4	1	4	3	4
ADM	0.2	0.17	0.25	0.33	0.5	0.5	0.5	0.25	1	3	2
TW	0.17	0.33	0.5	0.5	0.25	0.25	0.25	0.33	0.03	1	2
INT	0.5	0.5	1	0.25	0.33	0.25	0.5	0.25	0.5	0.5	1

Step 1: Construct the normalized decision matrix. This step transforms various attribute dimensions into non-dimensional attributes, which allows comparisons across criteria. The decision problem can be concisely expressed in the normalized decision matrix. Table 6.12 provides the normalized decision matrix.

Normalize scores or data as follows:

$$r_{ij} = X_{ij}/(\Sigma X^2_{ij}) \text{ for } i = 1, \ldots, m; \ j = 1, \ldots, n \qquad \text{Eq. 6.4}$$

TABLE 6.12

Normalized Decision Matrix

	PC	PDD	PP	RME	PPC	QC	SA	MGT	ADM	TW	INT	Weights
PC	0.0961	0.3846	0.2884	0.1923	0.0240	0.2884	0.1923	0.0317	0.4807	0.5769	0.1362	0.1362
PDD	0.0305	0.1219	0.3656	0.2437	0.0305	0.0402	0.2437	0.0402	0.7311	0.3656	0.2437	0.0839
PP	0.0438	0.0438	0.1329	0.6644	0.0332	0.0664	0.3986	0.0438	0.5315	0.2657	0.1329	0.0763
RME	0.0887	0.0887	0.0355	0.1774	0.0444	0.0585	0.1774	0.0444	0.5322	0.3548	0.7096	0.0553
PPC	0.3607	0.3607	0.3607	0.3607	0.0902	0.2705	0.3607	0.1803	0.1803	0.3607	0.2705	0.2058
QC	0.0360	0.3269	0.2179	0.3269	0.0360	0.1090	0.4359	0.3269	0.2179	0.4359	0.4359	0.1367
SA	0.0966	0.0966	0.0637	0.1932	0.0483	0.0483	0.1932	0.0483	0.3864	0.7727	0.3864	0.0511
MGT	0.2980	0.2980	0.2980	0.3973	0.0497	0.0328	0.3973	0.0993	0.3973	0.2980	0.3973	0.1470
ADM	0.0515	0.0438	0.0644	0.0851	0.1289	0.1289	0.1289	0.0644	0.2577	0.7732	0.5155	0.0437
TW	0.0692	0.1342	0.2034	0.2034	0.1017	0.1017	0.1017	0.1342	0.1342	0.4068	0.8136	0.0306
INT	0.2655	0.2655	0.5310	0.1328	0.1752	0.1328	0.2655	0.1328	0.2655	0.2655	0.5310	0.0334

Step 2: Construct the weighted normalized decision matrix. Not all of the selection criteria may be of equal importance and hence weighting was introduced from AHP technique to quantify the relative importance of the different selection criteria. The weighting decision matrix is simply constructed by multiply each element of each column of the normalized decision matrix by the random weights. Multiply each column of the normalized decision matrix by its associated weight. The weighted normalized decision matrix is shown in Table 6.13. An element of the new matrix is: $v_{ij} = w_j r_{ij}$

Step 3: Determine the ideal and negative ideal solutions. Tables 6.14 and 6.15 shows the weighted normalized decision matrix for ideal and non-ideal solutions, respectively.

Ideal solution, $A^* = \{v_1^*, \ldots, v_n^*\}$, where

$$v_j^* = \{\max(v_{ij}) \text{ if } j \in J; \ \min(v_{ij}) \text{ if } j \in J'\} \qquad \text{Eq. 6.5}$$

Negative ideal solution, $A' = \{v_{1'} \ldots, v_n'\}$, where

$$v' = \{\min(v_{ij}) \text{ if } j \in J; \ \max(v_{ij}) \text{ if } j \in J'\} \qquad \text{Eq. 6.6}$$

Step 4: Calculate the separation measures for each alternative. Tables 6.16a and 6.16b shows separation from ideal and non-ideal solutions respectively. The separation from the ideal alternative is:

$$S_i^* = \left[\Sigma \left(v_j^* - v_{ij}\right)^2\right]^{1/2} i = 1, \ldots, m \qquad \text{Eq. 6.7}$$

TABLE 6.13

Weighted Normalized Decision Matrix

	PC	PDD	PP	RME	PPC	QC	SA	MGT	ADM	TW	INT
PC	0.0131	0.0323	0.0220	0.0106	0.0049	0.0394	0.0098	0.0047	0.0210	0.0176	0.0064
PDD	0.0041	0.0102	0.0279	0.0135	0.0063	0.005	0.0124	0.0059	0.0319	0.0112	0.0081
PP	0.0060	0.0037	0.0101	0.0368	0.0068	0.0091	0.0204	0.0064	0.0232	0.0081	0.0044
RME	0.0121	0.0074	0.0027	0.0098	0.0091	0.0080	0.0091	0.0065	0.0232	0.0109	0.0237
PPC	0.0491	0.0303	0.0275	0.0200	0.0186	0.037	0.0184	0.0265	0.0079	0.0110	0.0090
QC	0.0049	0.0274	0.0166	0.0181	0.0074	0.0149	0.0223	0.0481	0.0095	0.0133	0.0146
SA	0.0132	0.0081	0.0049	0.0107	0.0099	0.0066	0.0099	0.0071	0.0169	0.0236	0.0129
MGT	0.0406	0.0250	0.0227	0.0220	0.0102	0.0045	0.0203	0.0146	0.0173	0.0091	0.0133
ADM	0.0070	0.0037	0.0049	0.0047	0.0265	0.0176	0.0066	0.0095	0.0113	0.0236	0.0172
TW	0.0094	0.0113	0.0155	0.0113	0.0209	0.0139	0.0052	0.0197	0.0059	0.0124	0.0272
INT	0.0362	0.0223	0.0405	0.0073	0.0361	0.0181	0.0136	0.0195	0.0116	0.0081	0.0177

TABLE 6.14

Weighted Normalized Decision Matrix for the Ideal Solution

	PC	PDD	PP	RME	PPC	QC	SA	MGT	ADM	TW	INT	Max
PC	0.0131	0.0323	0.0220	0.0106	0.0049	0.0394	0.0098	0.0047	0.0210	0.0176	0.0064	0.0394
PDD	0.0041	0.0102	0.0279	0.0135	0.0063	0.005	0.0124	0.0059	0.0319	0.0112	0.0081	0.0319
PP	0.0060	0.0037	0.0101	0.0368	0.0068	0.0091	0.0204	0.0064	0.0232	0.0081	0.0044	0.0368
RME	0.0121	0.0074	0.0027	0.0098	0.0091	0.0080	0.0091	0.0065	0.0232	0.0109	0.0237	0.0237
PPC	0.0491	0.0303	0.0275	0.0200	0.0186	0.037	0.0184	0.0265	0.0079	0.0110	0.0090	0.0491
QC	0.0049	0.0274	0.0166	0.0181	0.0074	0.0149	0.0223	0.0481	0.0095	0.0133	0.0146	0.0481
SA	0.0132	0.0081	0.0049	0.0107	0.0099	0.0066	0.0099	0.0071	0.0169	0.0236	0.0129	0.0236
MGT	0.0406	0.0250	0.0227	0.0220	0.0102	0.0045	0.0203	0.0146	0.0173	0.0091	0.0133	0.0406
ADM	0.0070	0.0037	0.0049	0.0047	0.0265	0.0176	0.0066	0.0095	0.0113	0.0236	0.0172	0.0265
TW	0.0094	0.0113	0.0155	0.0113	0.0209	0.0139	0.0052	0.0197	0.0059	0.0124	0.0272	0.0272
INT	0.0362	0.0223	0.0405	0.0073	0.0361	0.0181	0.0136	0.0195	0.0116	0.0081	0.0177	0.0405

TABLE 6.15

Weighted Normalized Decision Matrix for the Negative Ideal Solution

	PC	PDD	PP	RME	PPC	QC	SA	MGT	ADM	TW	INT	Min
PC	0.0131	0.0323	0.0220	0.0106	0.0049	0.0394	0.0098	0.0047	0.0210	0.0176	0.0064	0.0047
PDD	0.0041	0.0102	0.0279	0.0135	0.0063	0.005	0.0124	0.0059	0.0319	0.0112	0.0081	0.0041
PP	0.0060	0.0037	0.0101	0.0368	0.0068	0.0091	0.0204	0.0064	0.0232	0.0081	0.0044	0.0037
RME	0.0121	0.0074	0.0027	0.0098	0.0091	0.0080	0.0091	0.0065	0.0232	0.0109	0.0237	0.0027
PPC	0.0491	0.0303	0.0275	0.0200	0.0186	0.037	0.0184	0.0265	0.0079	0.0110	0.0090	0.0079
QC	0.0049	0.0274	0.0166	0.0181	0.0074	0.0149	0.0223	0.0481	0.0095	0.0133	0.0146	0.0049
SA	0.0132	0.0081	0.0049	0.0107	0.0099	0.0066	0.0099	0.0071	0.0169	0.0236	0.0129	0.0049
MGT	0.0406	0.0250	0.0227	0.0220	0.0102	0.0045	0.0203	0.0146	0.0173	0.0091	0.0133	0.0045
ADM	0.0070	0.0037	0.0049	0.0047	0.0265	0.0176	0.0066	0.0095	0.0113	0.0236	0.0172	0.0037
TW	0.0094	0.0113	0.0155	0.0113	0.0209	0.0139	0.0052	0.0197	0.0059	0.0124	0.0272	0.0052
INT	0.0362	0.0223	0.0405	0.0073	0.0361	0.0181	0.0136	0.0195	0.0116	0.0081	0.0177	0.0073

Similarly, the separation from the negative ideal alternative is:

$$S_i' = \left[\Sigma \left(v_j' - v_{ij} \right)^2 \right]^{1/2} i = 1, \ldots, m \qquad \text{Eq. 6.8}$$

Step 5: Calculate the relative closeness to the ideal solution C_i^*

$$C_i^* = S_i'/(S_i^* + S_i'), \ 0 < C_i^* < 1 \qquad \text{Eq. 6.9}$$

TABLE 6.16A

Separation from the Ideal Solution

	PC	PDD	PP	RME	PPC	QC	SA	MGT	ADM	TW	INT
PC	0.0007	0.0001	0.0008	0.0012	0.0003	0.0000	0.0009	0.0012	0.0003	0.0005	0.0011
PDD	0.0008	0.0005	0.0003	0.0007	0.0000	0.0007	0.0004	0.0007	0.0000	0.0004	0.0006
PP	0.0009	0.0000	0.0000	0.0009	0.0007	0.0008	0.0003	0.0009	0.0002	0.0008	0.0010
RME	0.0001	0.0003	0.0002	0.0002	0.0004	0.0002	0.0002	0.0003	0.0000	0.0002	0.0000
PPC	0.0000	0.0004	0.0009	0.0009	0.0005	0.0001	0.0009	0.0005	0.0017	0.0015	0.0016
QC	0.0019	0.0004	0.0009	0.0017	0.0010	0.0011	0.0007	0.0000	0.0015	0.0012	0.0011
SA	0.0001	0.0002	0.0002	0.0002	0.0004	0.0003	0.0002	0.0003	0.0000	0.0000	0.0001
MGT	0.0000	0.0002	0.0003	0.0009	0.0003	0.0013	0.0004	0.0007	0.0005	0.0010	0.0007
ADM	0.0004	0.0005	0.0005	0.0000	0.0005	0.0001	0.0004	0.0003	0.0002	0.0001	0.0001
TW	0.0003	0.0003	0.0003	0.0000	0.0001	0.0002	0.0005	0.0001	0.0005	0.0002	0.0000
INT	0.0000	0.0003	0.0011	0.0000	0.0000	0.0005	0.0007	0.0004	0.0008	0.0011	0.0005

Select the option with C_i^* closest to 1.

Table 6.17 presents the results of TOPSIS. The previous work has been validated through this tool as same factors as analyzed from questionnaire analysis, case studies, and AHP with maximum priority. So the factors selected for further validation are *production planning and control, product concept, quality control,* and *management.*

6.3. VIKOR Method

The compromise solution is a feasible solution that is the closest to the ideal solution, and a compromise means an agreement established by mutual concession. The compromise solution method, also known as VIKOR is the *Vlse Kriterijumska Optimizacija I Kompromisno Resenje* in Serbian, which means multi-criteria optimization (MCO), introduced as one applicable technique to implement within MADM. It focuses on ranking and selecting from a set of alternatives in the presence of conflicting criteria. The compromise solution, whose foundation was established is a feasible solution, which is the closest to the ideal, and here "compromise" means an agreement established by mutual concessions. The VIKOR method determines the compromise ranking list and the compromise solution by introducing the multi-criteria ranking index based on the particular measure of "closeness" to the "ideal" solution.

The procedure of VIKOR for ranking alternatives is as follows:

Step 1: Determine that best X_j^* and the worst X_j^- values of all criterion functions, where $j = 1, 2,.., n$. Table 6.18 shows the decision matrix

TABLE 6.16B

Separation from the Negative Ideal Solution

	PC	PDD	PP	RME	PPC	QC	SA	MGT	ADM	TW	INT
PC	0.000071	0.000762	0.000301	0.000036	0.000000	0.001209	0.000027	0.000000	0.000267	0.000168	0.000003
PDD	0.000000	0.000037	0.000564	0.000087	0.000004	0.000002	0.000009	0.000003	0.000771	0.000049	0.000016
PP	0.000005	0.000000	0.000042	0.001095	0.000010	0.000029	0.000278	0.000008	0.000381	0.000020	0.000001
RME	0.000088	0.000022	0.000000	0.000051	0.000041	0.000028	0.000040	0.000015	0.000421	0.000066	0.000441
PPC	0.001702	0.000501	0.000386	0.000146	0.000114	0.000847	0.000111	0.000348	0.000000	0.000010	0.000001
QC	0.000000	0.000507	0.000138	0.000174	0.000006	0.000100	0.000301	0.001864	0.000021	0.000071	0.000093
SA	0.000069	0.000010	0.000000	0.000034	0.000026	0.000003	0.000025	0.000005	0.000144	0.000352	0.000065
MGT	0.001304	0.000421	0.000334	0.000306	0.000033	0.000000	0.000250	0.000102	0.000166	0.000021	0.000077
ADM	0.000011	0.000000	0.000002	0.000001	0.000522	0.000194	0.000008	0.000034	0.000057	0.000399	0.000183
TW	0.000018	0.000037	0.000107	0.000037	0.000248	0.000076	0.000000	0.000212	0.000000	0.000053	0.000483
INT	0.000830	0.000223	0.001101	0.000000	0.000825	0.000117	0.000018	0.000148	0.000018	0.000001	0.000108

TABLE 6.17

Ranking from TOPSIS

Attributes	Si*	Si′	Si* + Si′	Ci=Si′/Si* + Si′	Rank
Product Concept	0.0731	0.0523	0.1254	0.4171	3
Product Design and Development	0.0763	0.0474	0.1237	0.3832	5
Process Planning	0.0739	0.0443	0.1182	0.3752	6
Raw Material and Equipment	0.0852	0.0502	0.1354	0.3706	7
Production Planning and Control	0.0724	0.064	0.1364	0.4694	1
Quality Control	0.0729	0.051	0.1239	0.4119	4
Strategic Agility	0.0819	0.0428	0.1247	0.343	8
Management	0.0648	0.0545	0.1193	0.4571	2
Administration	0.083	0.0348	0.1178	0.2954	11
Teamwork	0.083	0.0384	0.1214	0.316	9
Interpersonal	0.0746	0.0343	0.1089	0.3151	10

TABLE 6.18

Decision Matrix

	PC	PDD	PP	RME	PPC	QC	SA	MGT	ADM	TW	INT
PC	1	0.25	0.33	0.5	4	0.33	0.5	3	0.2	0.17	0.5
PDD	4	1	0.33	0.5	4	3	0.5	3	0.17	0.33	0.5
PP	3	3	1	0.2	4	2	0.33	3	0.25	0.5	1
RME	2	2	5	1	4	3	1	4	0.33	0.5	0.25
PPC	0.25	0.25	0.25	0.25	1	0.33	0.25	0.5	0.5	0.25	0.25
QC	3	0.33	0.5	0.33	3	1	0.25	0.33	0.5	0.25	0.5
SA	2	2	3	1	4	4	1	4	0.5	0.25	0.25
MGT	0.33	0.33	0.33	0.25	2	3	0.25	1	0.25	0.33	0.25
ADM	5	6	4	3	2	2	2	4	1	0.33	0.5
TW	6	3	2	2	4	4	4	3	3	1	0.5
INT	2	2	1	4	3	4	2	4	2	2	1

Step 2: Range Standardized Decision Matrix

$$X'_{ij} = [(X_{ij} - X_j^-)/(X_J^* - X_j^-)]$$ Eq. 6.10

Following Table 6.19 shows the range standardized decision matrix for the VIKOR method.

Step 3: Compute the S_i (the maximum utility) and R_i (the minimum regret) values, i = 1, 2,..., m by the relations:

TABLE 6.19

Range Standardized Decision Matrix

	PC	PDD	PP	RME	PPC	QC	SA	MGT	ADM	TW	INT
PC	0.1304	0.0000	0.0168	0.0789	1.0000	0.0000	0.0667	0.7275	0.0106	0.0000	0.3333
PDD	0.6522	0.1304	0.0168	0.0789	1.0000	0.7275	0.0667	0.7275	0.0000	0.0874	0.3333
PP	0.4783	0.4783	0.1579	0.0000	1.0000	0.4550	0.0213	0.7275	0.0283	0.1803	1.0000
RME	0.3043	0.3043	1.0000	0.2105	1.0000	0.7275	0.2000	1.0000	0.0565	0.1803	0.0000
PPC	0.0000	0.0000	0.0000	0.0132	0.0000	0.0000	0.0000	0.0463	0.1166	0.0437	0.1067
QC	0.4783	0.0139	0.0526	0.0342	0.6667	0.1826	0.0000	0.0000	0.1166	0.0437	0.0000
SA	0.3043	0.3043	0.5789	0.2105	1.0000	1.0000	0.2000	1.0000	0.1166	0.0437	0.3333
MGT	0.0139	0.0139	0.0168	0.0132	0.3333	0.7275	0.0000	0.1826	0.0283	0.0874	0.0000
ADM	0.8261	1.0000	0.7895	0.7368	0.3333	0.4550	0.4667	1.0000	0.2933	0.0874	0.3333
TW	1.0000	0.4783	0.3684	0.4737	1.0000	1.0000	1.0000	0.7275	1.0000	0.4536	0.3333
INT	0.3043	0.3043	0.1579	1.0000	0.6667	1.0000	0.4667	1.0000	0.6466	1.0000	1.0000

$$S_i = \sum W_j^* (X_j^* - X_{ij})/(X_j - X_j-) \qquad \text{Eq. 6.11}$$

$$R_i = \max \left[\sum W_j^* (X_j^* - X_{ij})/(X_j - X_j-) \right] \qquad \text{Eq. 6.12}$$

where W_j is the weight of the jth criterion which expresses the relative importance of criteria.

6.3.1. Coefficient of Variation

The weight of the criterion reflects its importance in MCDM. Range standardization was done to transform different scales and units among various criteria into common measurable units in order to compare their weights.

$$X'_{ij} = \left[(X_{ij} - \min X_{ij})/(\max X_{ij} - \min X_{ij}) \right]$$

$D' = (x')_{mxn}$ is the matrix after range standardization; $\max X_{ij}$, $\min X_{ij}$ are the maximum and the minimum values of the criterion (j) respectively, all values in D' are $(0 \leq x'_{ij} \leq 1)$. So, according to the normalized matrix. $D' = (x')_{mxn}$.

The standard deviation (σ_j) calculated for every criterion independently as:

$$\sigma_j = \sqrt{1/m} \sum (X'_{ij} - X^{-\prime}_j)^2 \qquad \text{Eq. 6.13}$$

Where X^{-}_j is the mean of the values of the jth criterion after normalisation and j= 1,2,....n.

After calculating (σ_j) for all criteria, the (CV) of the criterion (j) will be as shown

$$CV_j = \sigma_j / X^{-\prime}_j \qquad \text{Eq. 6.14}$$

The weight (W_j) of the criterion (j) can be defined as

$$W_j = CV_j / \sum CV_j \qquad \text{Eq. 6.15}$$

Where j = 1,2,....n.

Based on these, Table 6.20 provides the weights assigned to criteria.

Step 4: Compute the value $Q_i, i = 1, 2, \ldots, m$, by the relation

$$Q_i = \{v (S_i - S^*)/S^- - S^*)\} + \{(1 - v) (R_i - R^*)/R^- - R^*)\} \qquad \text{Eq. 6.16}$$

Where $S^* = \min S_i$, $S^- = \max S_i$, $R^* = \min R_i$, $R^- = \max R_i$, and v is the introduced weight of the strategy of S_i and R_i.

Step 5: Rank the alternatives, sorting by the S, R, and Q values in decreasing order. The results are shown in Table 6.21.

TABLE 6.20

Weights Assigned to Criteria

	X_j	σ	CV_j	W_j
PC	0.408379	0.305375	0.747774	0.067715
PDD	0.275257	0.289093	1.050265	0.095108
PP	0.28689	0.335672	1.170039	0.105954
RME	0.259091	0.320728	1.237896	0.112099
PPC	0.727273	0.34284	0.471405	0.042688
QC	0.570473	0.363753	0.637633	0.057741
SA	0.226182	0.295752	1.307584	0.118409
MGT	0.648997	0.371661	0.57267	0.051859
ADM	0.219403	0.304564	1.388152	0.125705
TW	0.200695	0.279016	1.390243	0.125895
INT	0.34303	0.366785	1.06925	0.096827

TABLE 6.21

Ranking List and Scores

Attributes	Si	Ri	Qi	Rank
Product Concept	0.320105	0.057741	0.238342	3
Product Design and Development	0.489607	0.096827	0.690176	5
Process Planning	0.474431	0.105954	0.725686	6
Raw Material and Equipment	0.540115	0.095108	0.744197	7
Production Planning and Control	0.20325	0.042688	0	1
Quality Control	0.440905	0.067715	0.453189	4
Strategic Agility	0.552163	0.112099	0.86165	8
Management	0.261585	0.051859	0.129431	2
Administration	0.595682	0.125895	1	11
Teamwork	0.576688	0.118409	0.930819	9
Interpersonal	0.58174	0.125705	0.981099	10

Recently, VIKOR has been widely applied for dealing with MCDM problems of various fields, such as environmental policy, data envelopment analysis, and personnel training selection. So the factors selected for further validation are *production planning and control, product concept, quality control*, and *management*.

6.4. Fuzzy Logic Using MATLAB

Fuzzy logic (FL) is based on the concept of a fuzzy set, which is without a crisp, clearly defined boundary. It contains elements having a partial degree of membership. A membership function (MF) defines mapping of each point in the input space to a membership value, which lies between 0 and 1. MF can have arbitrary curves whose shape can be define as a function that is simple, convenient, fast, and efficient. It defines the relationship between input and the output variables, thus, facilitating the optimization of process output. A single fuzzy if-then rule assumes the form "if x is A then y is B" where A and B are defined by fuzzy sets. The "if" part of the rule is called the premise, while the "then" part of the rule is called the conclusion. For example, "if the product concept is very low" and "quality is considerably high" then the result is "acceptable". The rules are determined through expert knowledge and are further refined following real life application and appraisal, which either confirm them or require them to be modified (Singh and Ahuja, 2012).

The function itself can be arbitrary curves whose shape can be define as a function that suits from the point of view of simplicity, convenience, speed, and efficiency. A function is a mathematical representation of the relationship between the input and output of a system or a process. It facilitates the optimization of process output by defining the true relationship between input and the output variables. Basically, it has been applied to validate the input factors determined from earlier tools. Therefore, only the identified factors are tested and results obtained justify the earlier obtained results. The fuzzy logic toolbox graphical user interface (GUI) tool to build a FIS is shown in Figure 6.1a.

Fuzzy Inference Systems

Fuzzy inference is the process of formulating the mapping from a given input to an output using FL. The mapping then provides a basis from which decisions can be made, or patterns discerned.

FISs have been successfully applied in fields such as automatic control, data classification, decision analysis, expert systems, and computer vision. Because of its multidisciplinary nature, FISs are associated with a number of names, such as fuzzy-rule-based systems, fuzzy expert systems, fuzzy modeling, fuzzy associative memory, FL controllers, and simply (and ambiguously) fuzzy systems. Figure 6.1b shows the procedure of FIS for the present study.

Fuzzification

The first step is to take the inputs and to determine the degree to which they belong to each of the appropriate fuzzy sets via MFs. In fuzzy logic

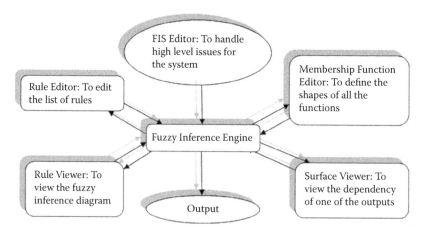

FIGURE 6.1
a: Tools Used in FL Toolbox.

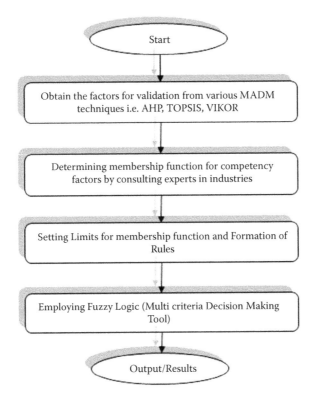

FIGURE 6.1
b: FIS procedure for present study.

toolbox software, the input is always a numerical value limited to the input variable and the output is a fuzzy degree of membership in the qualifying linguistic set (always the interval between 0 and 1).

Rule Evaluation

The FIS develops appropriate rules and on the basis of the rules the decision is made. This is principally established on the concepts of the fuzzy set theory, fuzzy "IF-THEN" rules, and fuzzy reasoning. FIS uses "IF..." and "THEN ..." statements, and the connectors existent in the rule statement are "OR" or "AND" to create the essential decision rules. The basic FIS can accept either fuzzy inputs or crisp inputs, but the outputs it provides are virtually all the time fuzzy sets.

Defuzzification

The input for the defuzzification process is a fuzzy set (the aggregate output fuzzy set) and the output is a single number. As much as fuzziness helps the rule evaluation during the intermediate steps, the final desired output for each variable is generally a single number.

6.4.1. Fuzzification

Figure 6.2 depicts the empirical transfer function as a FL system with inputs and output being fuzzified using appropriate MFs.

Here, the inputs are the factors like PC, PPC, QC, and MGT. The output is the result whose value shows whether to accept, under consider or reject the selection. The following sections narrate each component of the system.

6.4.2. Result: Checking Suitability

Product concept (PC) is measured by innovation or idea generation in a research and development cell. By consulting various experts from the industries, the fuzzy set rules defined for PC as: "If it is less or greater than 3% of the actual, then it is considered as low or high, or else it is optimum," as shown in Table 6.22. The transfer function in fuzzy is shown in Figure 6.3.

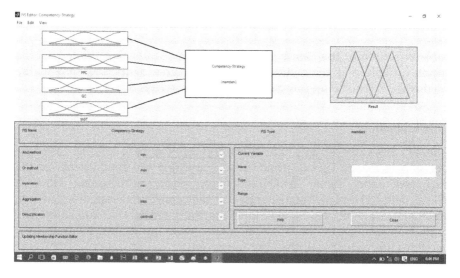

FIGURE 6.2
Empirical Transfer Function.

TABLE 6.22

Range for PC measurement

Fuzzy	Linguistic Value	Range
1	Low	-5 to -3
2	Optimum	-3 to 3
3	High	3 to 5

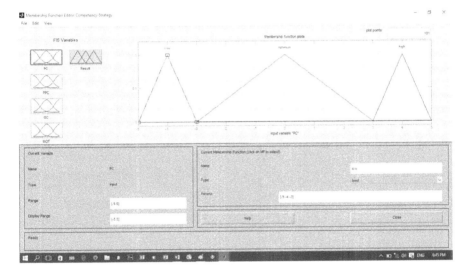

FIGURE 6.3
Transfer Function in Fuzzy Format of PC.

Production planning and control (PPC) measurement is by control plans required for production. By consulting various experts from the industries, the fuzzy set rules defined for PPC as: "If the control is less or greater than 2% of the actual, then it is considered as low or high, or else it is optimum," as shown in Table 6.23. The transfer function in fuzzy is shown in Figure 6.4.

Quality control (QC) is measured based on the defect rate of the delivered order quality. By consulting various experts from the industries, the fuzzy set rules defined for QC as: "If the actual quality is less or greater than 4% of the quality, then it is considered as low or high, or else it is optimum," as shown in Table 6.24. The transfer function in fuzzy is shown in Figure 6.5.

Management (MGT) is measured based on the support and cooperation from top management. By consulting various experts from the industries,

TABLE 6.23

Range for PPC Measurement

Fuzzy	Linguistic Value	Range
1	Low	-4 to -2
2	Optimum	-2 to 2
3	High	2 to 4

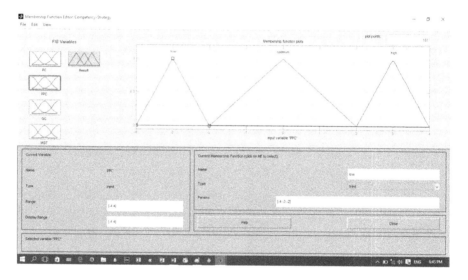

FIGURE 6.4
Transfer function in fuzzy format of PPC.

TABLE 6.24

Range for QC Measurement

Fuzzy	Linguistic Value	Range
1	Low	-6 to -4
2	Optimum	-4 to 4
3	High	4 to 6

the fuzzy set rules defined for MGT as: "If the support is less or greater than 3% of the required, then it is considered as low or high, or else it is optimum," as shown in Table 6.25. The transfer function in fuzzy is shown in Figure 6.6.

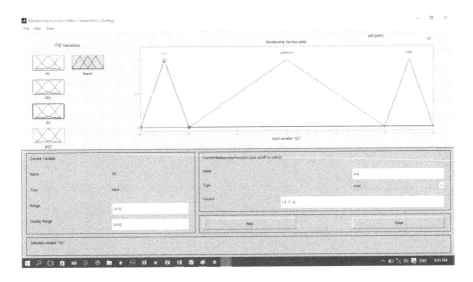

FIGURE 6.5
Transfer Function in Fuzzy Format of QC.

TABLE 6.25

Range for MGT Measurement

Fuzzy	Linguistic Value	Range
1	Low	-5 to -3
2	Optimum	-3 to 3
3	High	3 to 5

6.4.3. Fuzzy Evaluation Rules and Solution

The inputs here are the factors like product concept, production, planning and control, quality control, and management. Table 6.27 shows the formation of fuzzy rules. There are 81 rules following the format "if (condition a) and (condition b) and (condition c) and (condition d) then (result c)" corresponding to the combination of input conditions

The result value lies between 0 to 3 is considered as reject the system, between 3 to 6 is considered as poor (under consideration), between 6 to 8 is considered as acceptable, and between 8 to 10 is considered as an optimum system, as shown in Table 6.26. The transfer function in fuzzy is shown in Figure 6.7.

These if-then rule statements are used to formulate the conditional statements that comprise FL. For example, "if PC is high," "PPC required

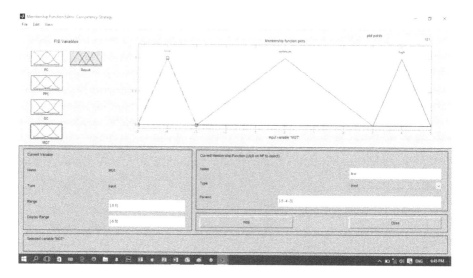

FIGURE 6.6
Transfer Function in Fuzzy Format of MGT.

TABLE 6.26

Range for Result Measurement

Fuzzy	Linguistic Value	Range
1	Reject	0–3
2	Under Consideration	3–6
3	Accept	6–8
4	Optimum	8–10

is high," "QC is optimum," and "MGT is low," then the result is "acceptable". These rules are formed with the expert knowledge, feedback, and guidance given by experts in the manufacturing industries and are further refined with experienced persons in the field of operation and production management from different manufacturing industries across India. The fuzzy set rules have been formed considering three different cases between of "PC, PPC, QC, MGT", when they are low, optimum, and high. Figures 6.8 and 6.9 shows various rules for the fuzzy logic.

A continuum of fuzzy solutions is presented in Figure 6.10a and b using the rule viewer of fuzzy toolbox of MATLAB. The inputs, i.e., PC, PPC, QC, and MGT can be set within the upper and lower specification limits and the output response is calculated as a score that can be translated into

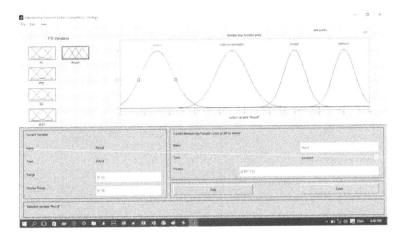

FIGURE 6.7
Transfer Function in Fuzzy Format of Results.

TABLE 6.27

Fuzzy Rules for Competency-Strategy

PC	PPC	QC	MGT
Low	Low	Low	Low
Optimum	Optimum	Optimum	Optimum
High	High	High	High

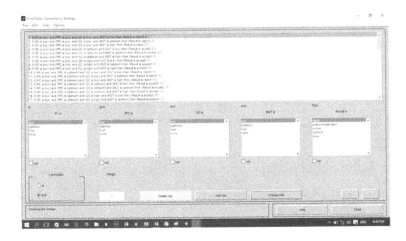

FIGURE 6.8
Fuzzy Rules.

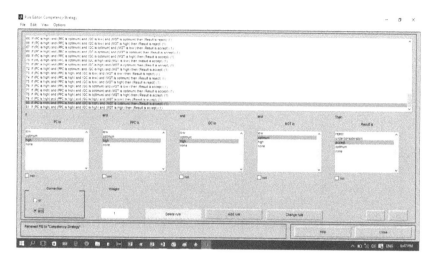

FIGURE 6.9
Fuzzy Rules.

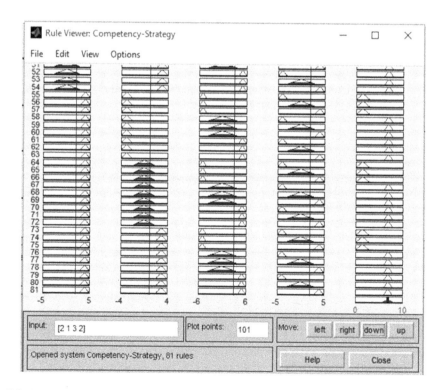

FIGURE 6.10
(a) Rule Viewer. (b) Rule Viewer.

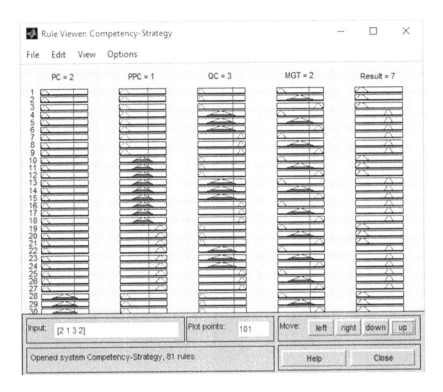

FIGURE 6.10
Continue.

linguistic terms. In this instance, if the value of PC is entered 2 (optimum), PPC 1 (optimum), QC 3 (optimum), and MGT 2 (optimum), the order output is 6, which specifies that the system is "acceptable" linguistically. Thus, from the analysis of the empirical data, case studies and qualitative techniques factors selected for further validation by using structural equation modeling (SEM) are *production planning and control, product concept, quality control*, and *management*.

7

Structural Equation Modeling

7.1. Validation of Qualitative Results through Structural Equation Modeling (SEM)

From empirical data obtained through the questionnaire survey, case studies, and AHP, validation of it has been done by using SEM. In this study, the input factors are *production, planning and control, product concept, quality control,* and *management*. Various output factors in this model are *sales, profit, growth and expansion, production capacity, production time, lead time, productivity, quality, competitiveness,* and *reliability*.

In this work, SEM has been carried out in Analysis of Moment Structures (AMOS) 21.0. It incorporates various statistical tools like causal modeling, confirmatory factor analysis (CFA), and path analysis. SEM evaluates, estimates, and specifies models of relationships between set of unobserved and observed variables. It inspects the interrelations through various equations that describe the relations among independent and dependent variables. Observed variables are the measured ones, designated graphically by using a rectangle or square. The responses, range from 1 to 4 and strongly disagree to strongly agree on a Likert scale. Unobserved variables are known as latent factors and are described graphically by ovals or circles.

Figure 7.1 shows a model of one outcome variable estimated by three observed variables. In AMOS, models are specified in terms of path diagrams, which are based on a set of standard rules. It is a significant skill for converting the data and theoretical hypotheses into a path diagram, which consists of:

The rectangle shows the observed variables (Ob1, Ob2, and Ob3) from the questionnaire.

The ellipse describes the latent variables (LVs) estimated from observed variables.

The error (e) shown in small circle, is an outcome in predicting a variable.

The single-headed arrow describes predictive relations.

The double-headed arrow shows covariance.

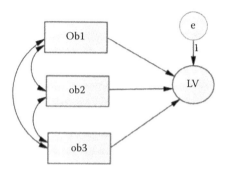

FIGURE 7.1
Basic SEM model.

The components of SEM: the outer model, which relates each observed variable to its corresponding indicators; and the inner model, which depicts the linear relation between the endogenous and exogenous unobserved variables. It has been described as a combination of exploratory factor analysis (EFA) and multiple regressions (Ullman, 2001). Various fit indexes are shown in Table 7.1.

CFA is used when there is some knowledge of the underlying estimated variable structure. EFA, in contrast to CFA, is considered for situations where the relation between the unobserved and observed variables is uncertain or unknown.

Based on empirical research, knowledge, or both, the relations that are proposed between variables and then this hypothesis is tested statistically. CFA rather than EFA is used for data analysis. The present research has been completed by CFA approach in SEM.

The core of SEM analysis indicates the fitness of the hypothesized model observed data and examination of the coefficients of hypothesized relationships. Generally, being fit means being consistent between two or more factors, and a good fit among factors will lead to better performance (Barrett, 2007; Byrne, 1989; Schumacker and Lomax, 2004). Several goodness-of-fit indicators have been used for assessing a model.

7.1.1. Variables Involved in the Study

There are number of variables that represent these models, but the selection of variables for this study was carefully made so that they should bear on the effectiveness of manufacturing competency and strategic success.

TABLE 7.1

Cut-off Criteria for Several Fit Indexes

Indexes	Shorthand	General for acceptable fit if data are continuous	Categorical data
Absolute/predictive fit chi-square	X^2	Ratio of x^2 to df ≤ 2 or 3, useful for nested models/model trimming	
Akaike information criterion	AIC	The smaller the better; good for model comparison (nonnested), not a single model	
Browne-cudeck criterion	BCC	The smaller the better; good for model comparison, not a single model	
Bayes information criterion	BIC	The smaller the better; good for model comparison (nonnested), not a single model	
Consistent AIC	CAIC	The smaller the better; good for model comparison (nonnested), not a single model	
Expected cross-validation index	ECVI	The smaller the better; good for model comparison (nonnested), not a single model	
Comparative fit		Comparison to a baseline (independence) or other model	
Normed fit index	NFI	≥ 0.95 for acceptance	
Incremental fit index	IFI	≥ 0.95 for acceptance	
Incremental fit index	TLI	≥ 0.95 can be $0 > TLI > 1$ for acceptance	0.96
Tucker-Lewis index	CFI	≥ 0.95 for acceptance	0.95
Relative noncentrality fit index	RNI	≥ 0.95 similar to CFI but can be negative, therefore CFI better choice	
Parsimonious fit			
Parsimony-adjusted NFI	PNFI	Very sensitive to model size	
Parsimony-adjusted CFI	PCFI	Sensitive to model size	
Parsimony-adjusted GFI	PGFI	Closer to 1 the better, thought typically lower than other indexes and sensitive to model size	
Other			
goodness-of-fit index	GFI	≥ 0.95 Not generally recommended	
Adjusted GFI	AGFI	≥ 0.95 Performance poor in simulation studies	
Hoelter .05 index		Critical N largest samples size for accepting that model is correct	
Hoelter .01 index		Hoelter suggestion, $N = 200$, better for satisfactory fit	

(Continued)

TABLE 7.1 (Cont.)

Indexes	Shorthand	General for acceptable fit if data are continuous	Categorical data
Root mean square residual	RMR	The smaller, the better; 0 indicate perfect fit	
Standardized RMR	SRMR	≤ 0.08	
Weighted root mean residual	WRMR	< 0.90	< 0.90
Root mean square error of approximation	RMSEA	< 0.06 to 0.08 with confidence interval	< 0.06

7.1.1.1. Dependent Variables

In this study, the dependent variables or the performance parameters have not been separated into different variables, but only into one that is, 'output' (OUTP), for depicting various firm performances of automobile manufacturing units, like growth and expansion, quality, reliability, cost, etc. This factor plays a significant role in leading the company successfully into a turbulent and competitive environment.

7.1.1.2. Independent Variables

Considering the variables, the manufacturing competency questionnaire was designed and sent to small, medium, and large-scale industries for data collection.

The variables are selected based on correlation, regression, and AHP analysis, as shown in Table 7.2.

7.2. SEM of the Manufacturing Competency Model

The questionnaire used was been divided into four parts. The first part was dedicated to gathering demographical information about the respondents and their respective organizations, such as name of the organization, position, job title, years of experience, etc. to ensure that the respondents had suitable backgrounds.

The questionnaire was also useful in identifying errors or discrepancies in responses received. Further, the second, third, and fourth part of the questionnaire measured the effectiveness of manufacturing competencies, strategic success, and output, respectively. Each statement in the questionnaire was designed to extract the respondent's opinion on the above parts

TABLE 7.2

Independent Variables Taken for the Study

Manufacturing Competency Independent Variables

Product Concept
Production Planning & Control
Quality Control
Management

in the context of organization's performance measurement using a four-point Likert scale. Figure 7.2 shows the framework of SEM_MC model.

7.2.1. Data Screening for Preliminary Analysis

After collecting the data by the questionnaire, various analysis techniques were applied for checking and increasing confidence in the data. This data helped in constructing a SEM_MC model in SEM using AMOS to deploy the interrelation between the variables in the study. SEM analysis finds the

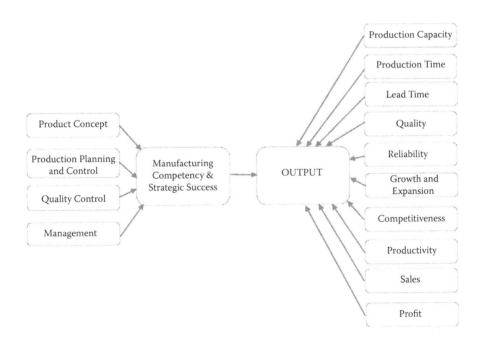

FIGURE 7.2
Framework of SEM_MC model.

variables that best explained the nature of the relationship between the variables and the model. The screening of the variables is a significant step as the requirement for this analysis is that the data is required to be multivariate normally distributed.

In this work, univariate normality test has been employed, as there is no method in SPSS or AMOS for testing the multivariate normality of the data. Skewness and kurtosis, as shown in Figure 7.3, are two major components for measuring data. Tables 7.3(a) and (b) shows kurtosis and skewness measures for independent and dependent variables.

As skewness relates to the symmetry of the distribution, a skewed variable is the one whose mean is not in the center of the distribution. Kurtosis, on the other hand, measures the spread of data relative to normal distribution and relates to the peakedness of the distribution (Pallant, 2005). According to Currie et al. (1999), the values of kurtosis ($<\pm 7$) and skewness ($<\pm 2$) are considered acceptable. Tables 7.3(a) and (b) measure the descriptive statistics of dependent and independent variables of SEM_MC model. The measures of skewness and kurtosis for all items are in range so the distribution of data does not depart from normality.

7.2.2. CFA

For measuring the suitability of data for CFA, the sample adequacy has been verified by using KMO (Kaiser-Meyer-Olkin) test and the correlations between various items have been verified by Bartlett's sphericity test (Pallant, 2005).

Bartlett's sphericity test should be calculated at significance level of p <0.05 and range for KMO index ranges from 0 to 1, for data to be suitable for CFA, 0.5 is considered as minimum value. Table 7.4 shows the measures of Bartlett's sphericity measures and KMO measures for independent and dependent variables and the values recommend that the data is appropriate for working with a CFA procedure.

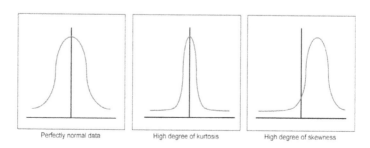

Perfectly normal data   High degree of kurtosis   High degree of skewness

FIGURE 7.3
Illustrating Kurtosis and Skewness.

TABLE 7.3(a)

Kurtosis and Skewness Measures for Independent Variables

						Skewness		Kurtosis	
Variables	Items	NStat.	MinStat.	Max Stat	Std.Dev. Stat.	Stat.	Std. Error	Stat.	Std. Error
	X11	118	1	4	0.8629	-0.79	0.223	-0.849	0.442
	X12	118	1	4	0.7976	0.132	0.223	-0.396	0.442
	X13	118	1	4	0.8051	-0.405	0.223	-0.736	0.442
PC	X14	118	1	4	0.7668	0.159	0.223	-0.714	0.442
	X15	118	1	4	0.7767	0.898	0.223	1.025	0.442
	X16	118	1	4	0.8310	0.466	0.223	-0.306	0.442
	X51	118	1	4	0.8439	0.090	0.223	-0.569	0.442
	X52	118	1	4	0.6877	-0.503	0.223	0.197	0.442
	X53	118	1	4	0.9018	0.613	0.223	-0.272	0.442
PPC	X54	118	1	4	0.5458	1.641	0.223	6.096	0.442
	X55	118	1	4	0.4517	1.476	0.223	0.927	0.442
	X56	118	1	4	0.9518	0.786	0.223	-0.207	0.442
	X57	118	1	4	0.8110	0.125	0.223	-0.431	0.442
	X61	118	1	4	0.8163	-0.799	0.223	0.513	0.442
	X62	118	1	4	0.8847	0.685	0.223	-0.049	0.442
QC	X63	118	1	4	0.5974	1.190	0.223	5.787	0.442
	X64	118	1	4	0.9598	0.118	0.223	-0.997	0.442
	X65	118	1	4	0.8119	0.586	0.223	-0.677	0.442
	X66	118	1	4	0.7312	0.231	0.223	-1.091	0.442
	X81	118	1	4	0.8020	-0.019	0.223	-0.844	0.442
	X82	118	1	4	0.7667	-0.056	0.223	-0.773	0.442
	X83	118	1	4	1.0079	0.010	0.223	-1.265	0.442
MGT	X84	118	1	4	0.8436	-0.262	0.223	-1.374	0.442
	X85	118	1	4	0.8149	0.195	0.223	-0.828	0.442
	X86	118	1	4	0.7507	-0.274	0.223	-1.173	0.442
	X87	118	1	4	0.8223	-0.168	0.223	-0.571	0.442

Figures 7.4–7.8 show the CFA for all the variables of SEM_MC model. Variables with regression weights below 0.7 should be removed, as the SEM model will be unfit because of these. So, as is seen from CFA, items X54 and X55 from PPC, item X15 from PC, item X82 from MGT, item X73 from QC, and item Y11 and Y12 from OUTP should be excluded from SEM_MC model.

After removal these items, Cronbach's alpha is used for reliability testing of the data. Table 7.5 shows Cronbach's alpha for all variables. They are acceptable as all the values are above 0.8.

TABLE 7.3(b)

Kurtosis and Skewness Measures for Dependent Variables

Variables	Items	N Statistic	Min Statistic	Max Statistic	Sted. Dev. Statistic	Skewness		Kurtosis	
						Statistic	Std. Error	Statistic	Sted. Error
	Y1	118	1	4	0.7641	-0.130	0.223	-1.267	0.442
	Y2	118	1	4	0.7603	-0.295	0.223	-0.778	0.442
	Y3	118	1	4	0.7900	0.032	0.223	-0.492	0.442
	Y4	118	1	4	0.7315	-0.675	0.223	0.432	0.442
	Y5	118	1	4	0.8963	-0.435	0.223	0.765	0.442
	Y6	118	1	4	0.9354	-0.042	0.223	-1.037	0.442
OUTP	Y7	118	1	4	0.8495	-0.205	0.223	-0.836	0.442
	Y8	118	1	4	0.7801	-0.776	0.223	-0.165	0.442
	Y9	118	1	4	0.8676	0.452	0.223	-0.919	0.442
	Y10	118	1	4	1.9762	0.34	0.223	- 0.987	0.442
	Y11	118	1	4	0.6542	0.138	0.223	- 0.098	0.442
	Y12	118	1	4	0.7795	0.295	0.223	-0.469	0.442

TABLE 7.4

Bartlett's Sphericity Measures and KMO Measures for Independent and Dependent Variables

Variable	KMO Measure	Bartlett's Sphericity Test	
		Chi-Square Value	P-Value
Product Concept	0.808	256.74	0.000
Production Planning and Control	0.743	298.793	0.000
Quality Control	0.814	364.750	0.000
Management	0.855	4820970	0.000
Output	0.823	830.154	0.000

7.2.3. SEM_MC Model and Result Analysis

Figure 7.9 shows the SEM_MC model in AMOS 21.0 for building the relation between variables involved in the research. The unstandardized SEM_MC model shows the regression coefficients that link the variables of this study. The output of this model provides the ordinary regression coefficient, covariance between independent variables, the error calculations of all variables, and the significance level for each

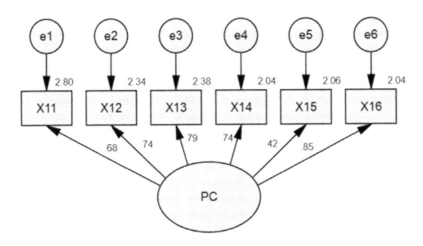

FIGURE 7.4
CFA Path Diagram for Product Concept Items.

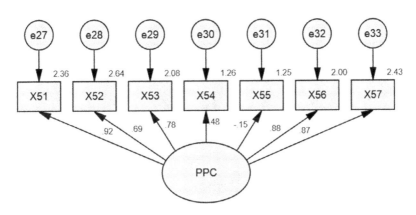

FIGURE 7.5
CFA Path Diagram for Production Planning and Control Items.

relation. The path analysis of between all variables and constructs is shown in the figure.

The output from first model was analyzed and compared with the criteria for cut-off for various fit indexes as shown in Table 7.6. The RMR (root mean square residual) value was analyzed as 0.48. It is the square root of the average squared amount of differences between estimates of sample variances and covariance. The GFI (goodness of fit

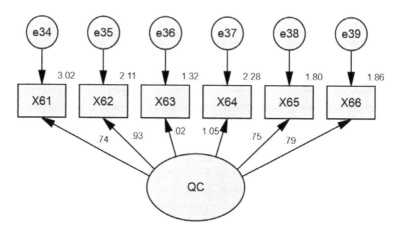

FIGURE 7.6
CFA Path Diagram for Quality Control Items.

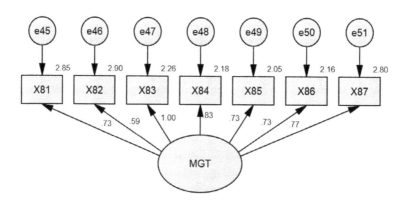

FIGURE 7.7
CFA Path Diagram for Management Items.

index) value was analyzed as 0.530. It calculates the ratio of variances that are accounted by estimation of covariance, though it is considered as an alternative to chi-square test (Jeong and Phillips, 2001). AGFI (adjusted goodness fit index) value was analyzed as 0.572. It is calculated according to degrees of freedom through saturated models. The model is considered a perfect fit, when the values of AGFI and GFI are near 0.95.

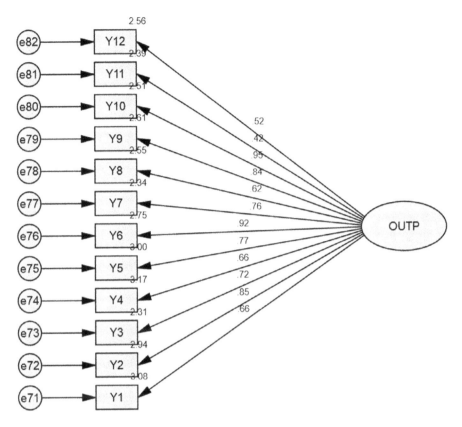

FIGURE 7.8
CFA Path Diagram for output Items.

7.2.4. Modification Indices for SEM_MC Model

The model was modified according to the modification indices, as shown in Table 7.6.

Modification indices indicate the improvements that may result by including a particular relation in the model. Modification indices are reduced to a small set, rather than showing all modifications and setting a threshold for these. The modification index is defined as an estimated amount for decrease in discrepancy function on repeating the analysis with the constraints on the removed parameter. AMOS, while displaying a modification index, also displays an estimated amount of change in parameter that can occur by removing constraints on it. The modified model after applying modification indices is shown in Figure 7.10.

TABLE 7.5

Cronbach's Alpha for all Variables of Model

Variables	Items	Cronbach's Alpha (α)
Product Concept	5	0.842
Production Planning & Control	5	0.824
Quality Control	5	0.883
Management	6	0.898
Output	10	0.905

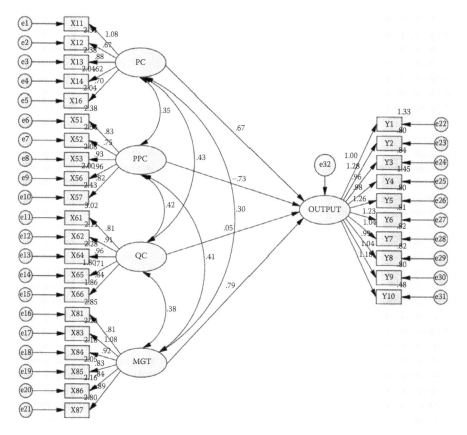

FIGURE 7.9
Model 1: SEM_MC model using AMOS 21.0.

TABLE 7.6

Modification Indices for Model

Covariance's of items			M.I	Par Change
e30	<->	e31	49.883	0.334
e27	<->	e31	14.787	0.183
e27	<->	e30	16.170	0.171
e26	<->	e28	17.062	-0.151
e22	<->	e31	15.731	-0.155
e8	<->	e19	57.336	0.255
e9	<->	e26	27.877	0.204
e11	<->	e18	34.799	0.163
e12	<->	e9	30.441	0.199
e1	<->	e31	20.178	0.156
e5	<->	e13	26.879	0.211
e5	<->	e4	23.653	0.200
e14	<->	e29	22.144	0.165
e10	<->	e6	13.345	0.121
e14	<->	e10	18.312	0.130
Regression Weights of the items			**M.I.**	**Par Change**
Y10	<—	QC	4.103	-0.211
Y10	<—	PC	5.315	-0.265
Y5	<—	PPC	4.320	0.188
Y5	<—	QC	4.016	0.167
X11	<—	MGT	5.885	-0.169
X12	<—	MGT	4.324	0.194
Y8	<—	PPC	6.060	0.220
Y8	<—	QC	5.248	0.188
Y3	<—	QC	4.095	-0.166
Y3	<—	PC	6.065	-0.222
X81	<—	PC	4.775	0.179
X83	<—	PC	4.150	-0.188
X13	<—	MGT	4.692	0.174

The comparison of model after applying the modification indices and before is shown in Table 7.7. It was seen that after deploying the modifications, there have been improvement in the model as RMR value decreased to 0.22, which is less as compared to earlier value. GFI value increased to 0.820, which is close to 1. The other values, required for making the model fit, related to SEM_MC model, is shown in Table 7.7.

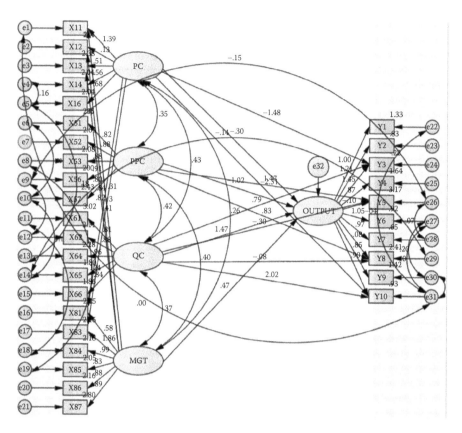

FIGURE 7.10
Model 2: Path Diagram of SEM_MC Model after Modification Index.

7.3. Competency-Strategy Model

Manufacturing competency has a significant effect on organizational performance. Competencies helps in attaining reliability, survival, profitability, market share, productivity, and even customer satisfaction, but these initiatives are directly affected by the cost, quality, and maintenance initiatives.

To validate this study empirically, the SEM_ MC Model was formed with SEM using AMOS 21.0. The data required for forming these models were obtained from respondents of different manufacturing organizations through the self-administered manufacturing competency questionnaire. Various important factors required for the SEM_MC model – i.e. production planning and control (PPC), management (MGT) product concept (PC), and quality control (QC) – were formulated from the questionnaire,

TABLE 7.7

Model Statistics

Model Fit Summary	Before Modification Indices	After Modification Indices	Recommended Value for Model Fit*
CMIN/Df	4.13	2.95	x2/ df<3.0
Degrees of freedom	427	399	Smaller is better
Probability level	0.00	0.00	
RMR	0.48	0.22	Smaller is better; 0 indicates perfect fit
Root-Mean-Square Error of Approximation (RNSEA)	0.163	0.098	< 0.08
Baseline Caparisons			
GFI	0.530	0.820	> 0.95
AGFI	0.562	0.842	> 0.95
Comparative Fit Index(CFI)	0.613	0.868	> 0.95
Incremental Fit Index (IFI)	0.617	0.879	> 0.95
Nomed Fit Index (NFI)	0.550	0.919	> 0.95
Relative Fit Index (RFI)	0.610	0.926	> 0.95
Tucker-Lewis Index (TLI)	0.579	0.837	> 0.95

case studies, analytical hierarchical processes, and from the extensive literature review. Further, various data examination techniques were applied on both the models. Through CFA, various items that were unfit were removed from independent and dependent variables. At last, using AMOS 21.0 software, modeling of manufacturing competency was carried out and the statistical data before and after the modification indices were compared. After comparing, the SEM_MC model was seen approaching near fit values, which implies that companies promoting competencies are improving business performances in terms of reliability, profitability, market share, productivity, quality of product, customer satisfaction, cost of product and maintenance initiatives, and hence this supports our previous work. Based on this, a competency strategy model has been developed.

Table 7.8 shows a generalized competency strategy model obtained from structural equation modeling. The values of standardized regression weights for various output factors as obtained from structural equation modeling are quality (0.75), production time (0.72), competitiveness (0.85), productivity (0.84), production capacity (0.77), reliability (0.77), growth and expansion (0.72), sales (0.77), profit (0.77), and lead time (0.72)

The developed model is based on selected input factors by using various qualitative and quantitative techniques such as AHP, TOPSIS, VIKOR, FUZZY, and SEM. The input factors from these analyses are product concept,

TABLE 7.8

Generalized Competency Strategy Model

PRODUCT CONCEPT	COMPETENCY STRATEGYMODEL	QUALITY
		PRODUCTION TIME
		COMPETITIVENESS
PRODUCTION PLANNING AND CONTROL		PRODUCTIVITY
QUALITY CONTROL		PRODUCTION CAPACITY
		RELIABILITY
MANAGEMENT		GROWTH AND EXPANSION
		SALES
		PROFIT
		LEAD TIME

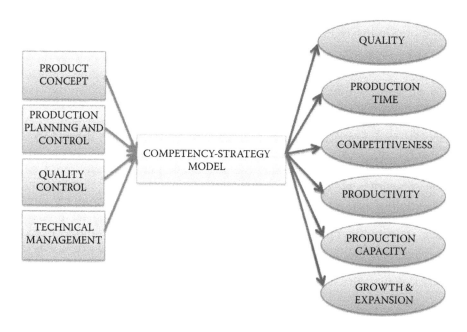

FIGURE 7.11
Finalized Competency Strategy Model.

production, planning and control, quality control, and management. The output factors are sales, profit, competitiveness, growth and expansion, quality, reliability, productivity, production capacity, production time, and lead time.

Figure 7.11 shows the final competency strategy model. This model has been developed based on the validation of selected input factors by using various qualitative and quantitative techniques, such as AHP, TOPSIS, VIKOR, FUZZY, and SEM.

8

Conclusions and Recommendations

In the previous chapters, data analysis and hypotheses testing were presented. This chapter starts by presenting the summary of this work followed by its contributions. Then, the limitations of the work are presented and, at finally, the scope for future works.

8.1. Summary of the Work

The work was aimed at identifying the attributes of manufacturing competency and strategic success of Indian manufacturing organizations in order to enhance performance. The study also critically examined manufacturing competency factors affecting the strategic success of these organizations. The major objective of this work was to examine, how manufacturing competency affected the strategy formation and thus organization's success. Lastly the work culminated with development of a manufacturing competency model for Indian automobile manufacturing units for growth and competitiveness. Primarily, the focus of this work was to determine the impact that manufacturing competency has on strategic success when viewed from an automobile manufacturing organization's perception.

In order for manufacturing competencies to have an impact on strategic success, it was important to include both human and technological interaction while developing a model. This chapter presents the conclusions and recommendations for the future from this study.

After reviewing more than 150 research papers, it was observed that impact of manufacturing competencies on strategic success has not been addressed yet. Based on the gaps from the literature, the objectives were framed and the issues for the study were listed along with the methodology adopted for this work. Initially there are 11 input factors with 61 sub variables and 12 output variables.

During this work, a questionnaire was prepared for the survey to be conducted in North Indian automobile manufacturing units. For the design of questionnaire, a pilot survey was created and finalized. During the pilot survey, academicians, industrialists, and the existing literature were consulted. Responses to the questionnaire were received from 118 units. For the survey, telephone interviews were made, e-mails and postages were sent, and in-person visits to the industrial units were made.

After completing the survey, a preliminary analysis of the data was carried out. Firstly, Cronbach alpha (a reliability analysis) was performed to analyze the questionnaire prepared. Then, various quantitative techniques were employed, like PPS and central tendency. This was followed by correlation and regression analysis of the above data. Based on the data analysis, factors were selected for detailed study in various manufacturing units.

Case studies were prepared in four different manufacturing units, as:

1. HMSI, Gurgaon (two-wheeler manufacturing units)
2. Suzuki, Manesar (four-wheeler manufacturing units)
3. SML Isuzu, Roopnagar (commercial vehicle manufacturing units)
4. Mahindra and Mahindra Ltd., Chandigarh (agricultural manufacturing units)

Data from each manufacturing unit was collected in detail by visiting the organizations in person. Letter of supports, duly signed by the above units, were received. (Appendix B).

From the case studies, it was shown that the performance of organizations has improved from the previous years. All the organizations have opted for new technologies and strategies in product development.

Based on the data collected from questionnaire survey and case studies, the synthesized results were obtained. Then results were followed by a qualitative analysis of the data. First, AHP followed by TOPSIS, VIKOR, and fuzzy logic were employed for parameter selection according to its importance and then for development of a model SEM was carried out by using AMOS 21.0 in SPSS 21.0 tool. Various important factors required for the SEM_MC model (i.e., quality control, production planning and control, product concept and management) were formulated from the questionnaire, case studies, and from the extensive literature review. Further various data analysis techniques like testing for kurtosis and skewness (i.e., for checking the normality variables) were applied on the model (i.e., SEM_MC). Through confirmatory factor analysis, various items which were negatively affecting the model were removed from the independent and dependent variables. During this, a model was developed from the results of the entire project.

8.2. Contribution of the Work

The work highlighted the attributes of manufacturing competencies and strategic success of Indian auto manufacturing organizations in order to enhance performance. The study also critically examined manufacturing competency factors affecting the strategic success of these organizations. Moreover, the study illustrated the relationship of manufacturing competencies and strategic success factors in Indian automobile manufacturing industries for overall business performance. The work critically examined the impact of manufacturing competency on strategy formation and thus the organization's success. Finally, the work culminated with the development of a manufacturing competency model for the Indian automobile industry for sustained strategic success. This study is a major contribution to the developing need of both academicians and professionals for better comprehension of the relationship of various manufacturing competencies and strategic success attributes.

The empirical analysis has been conducted in this work to evaluate the critical contribution of manufacturing competency factors in realizing strategic success in organizations. The association among manufacturing competence parameters and strategic success factors has been deployed to critically examine the impact of distinct manufacturing competency and strategic factors towards the fulfillment of organizational sustainability objectives. In this admiration, it formulates valuable ramifications for managers with respect to the technique development. The conclusions drawn from the work have been highlighted below.

8.2.1. PPS Results

The work provides an insight into exploits of Indian entrepreneurs regarding manufacturing competency and strategic success and provides an assessment of prevailing status of Indian entrepreneurs regarding different competency and strategy parameters:

a. The analysis of significant attributes of major product concept (idea generation) issues revealed that a significantly large number of organizations have evolved well-planned and structured concept generation process (percent point scored, PPS=70), promote innovation and marketing (PPS=58.4), encouragement for inter departmental teams (PPS=51.0), centralized planning (PPS=51.0), and creativity (PPS=58.0).

b. The analysis of significant attributes of product design and development issues revealed that a significantly large number of organizations have an effective design technology (PPS=67.8), computer technology for analysis (PPS=54.0), product life cycle (PPS=5.2), aesthetics and ergonomics of products (PPS=57.8), and simulation and modeling (PPS=53.8).

c. The analysis of issues related to process planning showed that many manufacturing organizations have an effective process planning program (PPS=67.16), tracking process planning costs (PPS=7.7), material and machine selection (PPS=56.1), group technology (PPS=5.8), and finishing and assembly of products (PPS=63.7).

d. A close analysis of various issues related to raw material and equipment has indicated that many manufacturing organizations use ERP software for record keeping (PPS=68.0), whereas some other factors like having own transportation (PPS=5.2), inventory storage (PPS=58.2), and different departments involvement for machine selection (PPS=55.7) need immediate attention, since these factors have been found to be underperforming.

e. The analysis of data obtained from the survey has indicated that many manufacturing organizations have generally reported low performance regarding production, planning, and control factors. The results revealed that many manufacturing organizations concentrate on precision and accuracy (PPS=65.9), green manufacturing (PPS=52.1), hydraulic and pneumatic system (PPS=5.0), and computer-aided manufacturing (PPS-59.1).

f. The analysis of significant attributes of quality revealed that many manufacturing organizations have shown acceptable performance rating. The data has indicated that that many manufacturing organizations test products under actual conditions (PPS=75.4), technology for quality analysis (PPS=57.0), and life cycle analysis (PPS=52.7).

g. A reasonable rating has been observed from the survey regarding strategic agility attributes. The results revealed that many manufacturing organizations have exhibited a strong desire to realize various strategic success issues, quality conformance (PPS=66.9), a strong customer base (PPS=62.3), competitive pricing (PPS=65.0), market shares (PPS=71.4), and profit (PPS=64.0).

h. The significant attributes of management depicted that many manufacturing organizations have better production planning and control functions (PPS=72.4), coordination between departments (PPS=7.0), and enhanced production capabilities (PPS=71.2); but some improvement can be suggested for other factors, like efficient information flow (PPS=56.6), information analysis in departments (PPS=54.4), crises management (PPS=54.0), and risk management (PPS=51.3), since these factors have been underperforming.

i. The analysis of data obtained from the survey has indicated that many manufacturing organizations have shown unsatisfactory performance (PPS) regarding team work factors. The results reveal that many manufacturing organizations emphasize on improving communication and coordination among team members (PPS=68.8), coordinated efforts for fostering next-generation technology (PPS=62.1), and the better promotion of products (PPS=68.2), whereas some other factors need immediate attention since these factors have been found to be underperforming.

j. The analysis of significant attributes of administration has indicated that many manufacturing organizations have efficient administration and management (PPS=71.6), top level management commitment (PPS=63.1), policy formation (PPS=67.6), and support and encouragement from top management (PPS=56.3).

k. The analysis of significant attributes of interpersonal issues has revealed that many manufacturing organizations have worked aggressively for building up the self-confidence of employees (PPS=71.6), safety and health awareness (PPS=58.7), self-managed teams (PPS=57.6), employee empowerment (PPS=56.8), a highly skilled workforce (PPS=55.0), and stress management (PPS=52.3).

l. The analysis of significant attributes of organizations has revealed that many manufacturing organizations have revealed acceptable performance (PPS) regarding major attributes. The organizations work on quality (PPS=79.2), production capacity (PPS=77.0), reliability (PPS=75), production time (PPS=73.5), productivity (PPS=68.6), competitiveness (PPS=63.7), sales (Annually) (PPS=65.2), a strong customer base (PPS=64.0), and profit (annually) (PPS=62.7). However, there is an emergent need to affect improvements in lead time (PPS=57.6), growth and expansion (PPS=58.4), and market share (PPS=59.7), since these factors have been underperforming.

8.2.2. Correlation Results

The work has highlighted contributions of various manufacturing competency attributes in the Indian manufacturing industry for realizing strategic success in a fiercely competitive marketplace. The data analysis has been conducted in this work to evaluate the impact of manufacturing competencies in realizing strategic success in various organizations. The association among manufacturing competence parameters and strategic success factors has been deployed to critically examine the impact of distinct manufacturing competency and strategy factors towards the fulfillment of organizational sustainability objectives. Thus manufacturing managers must develop competencies for managing manufacturing initiatives efficiently for realizing global strategic success. Therefore, the effective adoption of manufacturing competency factors contributes towards the realization of sustained strategic success.

a. The analysis of correlation matrix has revealed that the correlations obtained between the product concept and output were significant and affected the organization in a positive manner. The correlation of the product concept with the parameters – i.e., production capacity (r = 0.606), production time (r = 0.613), lead time (r = 0.450), quality (r = 0.675), reliability (r = 0.727), productivity (r = 0.500), growth and expansion (r = 0.527), competitiveness (r = 0.684), sales (r = 0.404), profit (r = 0.453), market share (r = 0.448), and customer base (r = 0.348) – was significantly positive.

b. The analysis of the correlation matrix showed that the correlations obtained between product design and development and all process of output were significant and affected the organization in a positive manner. The correlation of the product design and development with the parameters – i.e., production capacity (r = 0.606), production time (r = 0.635), lead time (r = 0.527), quality (r = 0.692), reliability (r = 0.758), productivity (r = 0.554), growth and expansion (r = 0.556), competitiveness (r = 0.696), sales (r = 0.472), profit (r = 0.505), market share (r = 0.471), and customer base (r = 0.436) – was significantly positive.

c. The analysis of the correlation matrix showed that the correlations obtained between process planning and all process of output were significant and affected the organization in a positive manner. The correlation of the process planning with the parameters – i.e., production capacity (r = 0.550), production time (r = 0.469), lead time (r = 0.371), quality (r = 0.595), reliability (r = 0.746), productivity (r = 0.468), growth and expansion (r = 0.472), competitiveness (r = 0.675), sales (r = 0.450), profit (r = 0.451), market share (r = 0.354), and customer base (r = 0.359) – was significantly positive.

d. The analysis of correlation matrix revealed that the correlations obtained between the raw material and equipment and all output parameters were significant and influenced the organization in a positive manner. The correlation of the raw material and equipment with the parameters – i.e., production capacity (r = 0.672), production time (r = 0.631), lead time (r = 0.451), quality (r = 0.690), reliability (r = 0.770), productivity (r = 0.553), growth and expansion (r = 0.552), competitiveness (r = 0.704), sales (r = 0.478), profit (r = 0.511), market share (r = 0.526), and customer base (r = 0.412) – was significantly positive.

e. The analysis of the correlation matrix showed that the correlations obtained between the production planning and all process of output were significant and affected the organization in a positive manner. The correlation of the production planning with the parameters – i.e., production capacity (r = 0.614), production time (r = 0.549), lead time (r = 0.384), quality (r = 0.590), reliability (r = 0.688), productivity (r = 0.496), growth and expansion (r = 0.553), competitiveness (r = 0.638), sales (r = 0.443), profit (r = 0.422), market share (r = 0.508) and customer base (r = 0.481) – was significantly positive.

f. The analysis of the correlation matrix showed that the correlations obtained between the quality control and all process of output were significant and affected the organization in a positive manner. The correlation of the quality control with the parameters – i.e., production capacity ($r = 0.653$), production time ($r = 0.628$), lead time ($r = 0.427$), quality ($r = 0.586$), reliability ($r = 0.652$), productivity ($r = 0.564$), growth and expansion ($r = 0.531$), competitiveness ($r = 0.525$), sales ($r = 0.395$), profit ($r = 0.377$), market share ($r = 0.358$), and customer base ($r = 0.453$) – was significantly positive.

g. The analysis of the correlation matrix showed that the correlations obtained between the strategic agility and all process of output were significant and affected the organization in a positive manner. The correlation of the strategic agility with the parameters – i.e., production capacity ($r = 0.679$), production time ($r = 0.677$), lead time ($r = 0.506$), quality ($r = 0.709$), reliability ($r = 0.643$), productivity ($r = 0.612$), growth and expansion ($r = 0.590$), competitiveness ($r = 0.569$), sales ($r = 0.547$), profit ($r = 0.554$), market share ($r = 0.482$), and customer base ($r = 0.474$) – was significantly positive.

h. The analysis of the correlation matrix showed that the correlations obtained between the management and all process of output were significant and affected the organization in a positive manner. The correlation of the management with the parameters – i.e., production capacity ($r = 0.516$), production time ($r = 0.640$), lead time ($r = 0.597$), quality ($r = 0.547$), reliability ($r = 0.700$), productivity ($r = 0.581$), growth and expansion ($r = 0.606$), competitiveness ($r = 0.639$), sales ($r = 0.529$), profit ($r = 0.655$), market share ($r = 0.521$), and customer base ($r = 0.515$) – was significant.

i. The analysis of the correlation matrix showed that the correlations obtained between the team work and all process of output were significant and affected the organization in a positive manner. The correlation of the teamwork with the parameters – i.e., production capacity ($r = 0.607$), production time ($r = 0.652$), lead time ($r = 0.494$), quality ($r = 0.647$), reliability ($r = 0.806$), productivity ($r = 0.596$), growth and expansion ($r = 0.555$), competitiveness ($r = 0.708$), sales ($r = 0.522$), profit ($r = 0.584$), market share ($r = 0.508$), and customer base ($r = 0.467$) – was significantly positive.

j. The analysis of the correlation matrix showed that the correlations obtained between the administration and all process of output were significant and affected the organization in a positive manner. The correlation of the administration with the parameters – i.e., production capacity ($r = 0.527$), production time ($r = 0.566$), lead time ($r = 0.480$), quality ($r = 0.624$), reliability ($r = 0.685$), productivity ($r = 0.575$), growth and expansion ($r = 0.482$), competitiveness ($r =$

0.631), sales (r = 0.491), profit (r = 0.507), market share (r = 0.413), and customer base (r = 0.488) – was significantly positive.

k. The analysis of the correlation matrix showed that the correlations obtained between the interpersonal and all process of output were significant and affected the organization in a positive manner. The correlation of the interpersonal with the parameters – i.e., production capacity (r = 0.528), production time (r = 0.554), lead time (r = 0.450), quality (r = 0.553), reliability (r = 0.738), productivity (r = 0.470), growth and expansion (r = 0.449), competitiveness (r = 0.610), sales (r = 0.400), profit (r = 0.432), market share (r = 0.475), and customer base (r = 0.498) – was significant.

8.2.3. Regression Results

The work has been extended further to establish a mathematical model between the dependent and independent variables involving manufacturing competencies and strategic success attributes. For this purpose, ANOVA analysis and t-tests have been performed.

a. The key predictors identified from the regression analysis of *production capacity* were product concept, raw material and equipment, production planning and control, quality control attributes of manufacturing competencies, strategic agility, and management attributes of strategic success.

b. The predictors identified from the regression analysis of *production time* were product concept, product design and development, process planning, production planning and control, quality control parameters of manufacturing competencies, strategic agility, management, and teamwork parameters of strategic success.

c. The predictors identified from the regression analysis of *lead time* was product concept, product design and development, raw material and equipment, process planning and control, quality control parameters of manufacturing competencies, and management parameters of strategic success.

d. The predictors identified from the regression analysis of *quality* were product concept, production planning and control and quality control parameters of manufacturing competency, and management and administration parameters of strategic success.

e. The predictors identified from the regression analysis of *reliability* were product concept, product design and development, raw material and equipment, production planning and control, quality control parameters of manufacturing competencies, and management and interpersonal parameters of the strategic success.

f. The predictors identified from the regression analysis of *productivity* included product concept; process planning; production planning and control; quality control parameters of manufacturing competencies; and management, administration, and interpersonal parameters of strategic success.

g. The predictors identified from the regression analysis of *growth and expansion* were product concept, product design and development, production planning and control, quality control parameters of manufacturing competency, and management and administration parameters of the strategic success.

h. The key predictors identified from the regression analysis of *competitiveness* were product concept, product design and development, production planning and control, quality control parameters of manufacturing competencies, and management and administration parameters of the strategic success.

i. The key predictors identified from the regression analysis of *sales* were product concept, production planning and control, quality control parameters of manufacturing competency, and management and interpersonal parameters of the strategic success.

j. The predictors identified from the regression analysis of *profit* were product concept; raw material and equipment; production planning and control; quality control parameters of the manufacturing competencies; and management, administration, and interpersonal parameters of the strategic success.

k. The key predictors identified from the regression analysis of *market share* were product concept, process planning, production planning and control, quality control parameters of manufacturing competencies, and management and teamwork parameters of strategic success.

l. The predictors identified from the regression analysis of *customer base* was product concept, production planning and control, and quality control parameters of manufacturing competencies and management parameters of strategic success.

8.2.4. Results of Preliminary Data Analysis

The empirical analysis has highlighted that product concept, production, planning and control, quality control, and management have emerged as significant contributors for realizing sustained competitiveness in the organization.

a) Product concept has a great impact on an organization's performance as it is related to the research and development department of the organization, thus involving innovation and idea generation.

b) Production, planning, and control plays an important role in the success of an organization, as it involves production, which directly deals with manufacturing, planning, which relates to plan formed for manufacturing and control refers to the secondary plans or controlling operations for the manufacturing process.

c) Quality control directly affects the market value of an organization, as better quality products lead to improved sales, thus a better market value and performance in the organization.

d) Management support and commitment significantly affects the success of the organization, since management initiatives provide a much-needed impetus for motivating employees towards organizational goals and also provide resources and investments, thereby leading to growth of the organization.

8.2.5. Case Studies Results

Further, the detailed, multi descriptive case studies were conducted in selected automobile manufacturing organizations across the northern part of India and have made serious interventions regarding manufacturing competency and strategic success. The following key issues have been highlighted through the case study:

a) The case studies have depicted organizations that have deployed various manufacturing competencies for attaining strategic success, thereby evolving significant production system innovations successfully, attaining cost-effective production capabilities, mitigating production wastages, and evolving and aligning personnel competencies with organizational sustainability objectives through empowerment and facilitation.

b) The manufacturing organization has been successful in creating a congenial atmosphere for implementation of manufacturing competencies through establishing improved communication in the organization, ensuring enhanced employee involvement, motivation, and empowerment; and by putting in resources for improving the employee competencies through training and skill-building.

c) Finally, an evaluation of achievements accrued through competencies has been highlighted to evolve the consensus on strategic success of an organization in a highly competitive environment. The study reveals that companies have evolved from past experiences by introducing innovative strategies and technologies in their products. Thus, to be able ensure the realization of sustained strategic success, the organizations must consistently continue to foster manufacturing competencies.

8.2.6. Qualitative Analysis Results

For parameter selection from the questionnaire, analysis has been done using an analytical hierarchy process (AHP). A consistency ratio has been computed as the ratio of the consistency index and a random consistency index. For the study, the value of CR that has been obtained is less than 0.1 (9.4%), which means the judgments considered for the study have been consistent and acceptable. Further, to validate the results AHP, TOPSIS, VIKOR, and fuzzy logic have been employed and similar results have been obtained. For further validation of results obtained from qualitative techniques, a structural equation modeling analysis has been conducted. This study uses the confirmatory factor analysis approach using structural equation modeling in AMOS 21.0 software to employ the relationship between manufacturing competencies and strategic success in the study.

a) A model fit summary of a SEM_MC model RMR value was 0.48. Similarly, the goodness of fit index (GFI) value was 0.530. The normed fit index (NFI) value was 0.550

b) The model fit summary of a SEM_MC model after deploying modification indices has indicated that the RMR value was observed 0.22, and also the GFI for the model was observed at 0.82, which indicated that the model after modifications does provide a better fit with respect to the NFI, which also was observed at 0.919. Thus, the SEM study confirms and validates the SEM_MC model.

8.2.7. Competency-Strategy Model

Finally, the work culminated with the development of competency-strategy model. The validation of the model confirmed the factors that emerged as significant. The work presented the evolution of a manufacturing competencies and a strategic success model in the Indian automobile manufacturing industry. The model has been based upon an extensive literature review, from analysis of the manufacturing competency questionnaires and results from case studies conducted in selected automobile manufacturing organizations.

8.3. Major Findings of the Study

This work is a significant contribution to the developing needs of both academicians and professionals for understanding the relationship of various manufacturing competencies and strategic success attributes. The conclusions drawn from this study are:

1. Product concept, production planning and control, quality control, and technical management have emerged as significant contributors to sustained competitiveness in the organization.

2. Product concept has a significant impact on an organization's competency as it involves innovation, development, and implementation.

3. Production planning and control plays a vital role in enhancing manufacturing competency, as it directly deals with manufacturing strategy planning, execution, and controlling the operations. Its contribution to strategic success is higher than that of product concept.

4. Quality control directly affects the market value of a product and contributes significantly to the strategic success of an organization.

5. Strong management support provides the much-needed impetus for motivating employees towards organizational goals by making the required resources available, which ultimately leads to the success and growth of any organization.

6. The qualitative competency-strategy model developed for the Indian automobile industry has highlighted the significant contributions of manufacturing competency factors to the strategic success of an organization.

8.4. Limitations of the Work

The limitations of the present work presents suggestions for future studies. The samples in this study were collected from automobile organizations. The validity of the findings regarding the relationship between manufacturing competencies and organization performance may be hampered, as data on manufacturing practices and organizational performance were collected around the same point of time. At last, organization performance may be affected by other variables not accounted for in this work. It would be useful to examine the organization performance by taking legal and economic factors into account. Further limitations of the study are:

1. The work concentrated on manufacturing competencies only. Other competencies, such as product design and development, maintenance, supply chain, green manufacturing, and a few others can also affect the organization's performance.

2. This study has been conducted in the automobile manufacturing industry only. Factors may vary according to manufacturing industries of different other products like bicycles, machines and machine tools, material handling equipment, and farm and agri-machinery.

3. In this work, only four significant factors (product concept, production planning and control, quality control, and management support) were considered. Other factors (like design and development, process planning, and total productive maintenance) could also have a significant effect on firm performance.

4. The scope of this work has been limited to Northern India only. The significance of issues may differ in other parts, such as the south or central, or nationwide.

5. A qualitative competency-strategy model has been developed. No other modeling techniques have been explored in this work.

8.5. Suggestions for Future Work

The primary aim of this work is to synthesis strategic success concepts and explore manufacturing competencies for automobile manufacturing organizations, while a similar study can also be conducted in the future for other Indian product, process, and service industries.

The work is aimed at developing manufacturing competencies and a strategic success model for the North Indian automobile and auto parts manufacturing organizations. Various manufacturers have been treated alike, irrespective of the sector of manufacturing organizations. Another direction for future work is developing area-wise, sector-wise, and product-wise competency models for the automobile manufacturing industry. Thus, individual case studies could be conducted for different areas, products, and sectors of the greater manufacturing industry, and accordingly the typical methodologies can also be evolved in future.

While this work provides an insight into manufacturing competencies and their relation with strategic success, it has also discovered areas that could improve from further work. This work focused only on manufacturing competencies of automobile organizations. Future works could focus on other competencies as well. By doing so, a better and broader understanding of the effects that other competencies have on organization's performance may be accomplished.

1. Manufacturing competencies of automobile manufacturing units have been explored. Future work could focus on the other competencies, like innovation, design and development, supply chain, green manufacturing, and such others.

2. Various other factors like process planning, strategic agility, teamwork, and goal orientation may be considered for future studies. They may also have an impact on organization's strategic success.

3. Future work could also concentrate on sectors like large-scale, medium-scale, and small-scale industries for competency model development in those industries.

4. Such studies can also be conducted in other regions of India to develop a holistic competency model for the entire Indian automobile manufacturing industry.

5. A mathematical model for depicting competency and strategy relations could be attempted in future, as in the present study only a qualitative model has been developed.

Appendix – A

QUESTIONNAIRE

for

PhD

RESEARCH WORK

IMPACT OF MANUFACTURING COMPETENCY

ON STRATEGIC SUCCESS OF THE AUTOMOBILE INDUSTRY

By

CHANDAN DEEP SINGH

ASSISTANT PROFESSOR

DEPARTMENT OF MECHANICAL ENGINEERING

DEPARTMENT OF MECHANICAL ENGINEERING

PUNJABI UNIVERSITY

PATIALA – 147002

MAILING ADDRESS
Please mail the questionnaire at the following address:
CHANDAN DEEP SINGH
S/o Prof. BHAWDEEP SINGH TANGHI
#20778-B (TANGHI-VILLA), St. No. 25-A,
AJIT ROAD, BATHINDA – 151001
E-Mail: er.chandandeep@gmail.com, chandandeep@pbi.ac.in
Contact Nos.: 98766-18111(M), 081469–21111(M)

MANUFACTURING COMPETENCY QUESTIONNAIRE
GENERAL ORGANIZATIONAL INFORMATION

Organization's Name				
Organization's Address				
Respondent's Name and Designation				
Respondent's E-Mail Address				
Respondent's Contact No./Fax No.				
i.	PRODUCTS OF THE ORGANIZATION (Please Specify)				
ii.	PRESENT TURNOVER (In Crores of Rupees)	<10 Crores	10–50 Crores	50–100 Crores	>100 Crores
iii.	NUMBER OF EMPLOYEES	<200	201–500	501–1,000	>1,000
iv.	MARKET SHARE	<20%	20–40%	40–60%	>60%

(Please tick the appropriate choice)

MANUFACTURING COMPETENCY					
S. No.	A – PRODUCT CONCEPT (IDEA GENERATION)				
1	Do you have a well-planned and structured concept generation process in your organization?	Not at all	To some extent planned	Partially planned	Well planned
2	Do the company policies promote innovation?	Not at all	To some extent	Reasonably	To a great extent
3	Do you feel that the marketing department is adequately motivated to get an idea about the new product?	Not at all	To some extent	Reasonably	To a great extent

4	Does the organization encourage the deployment of inter departmental teams to identify and create new ideas?	Not at all	To some extent	Reasonably	To a great extent
5	Is the organization flexible enough for making changes during operations and maintenance to satisfy customer needs?	Not at all	To some extent	Reasonably well	To a great extent
6	Does the organization use a centralized or decentralized planning structure for idea generation?	No organization at all	Centralized organization	Decentralized organization	Uses both organizations

B – PRODUCT DESIGN AND DEVELOPMENT

1	Does your organization have an effective design technology program (CAD)?	Not at all	To some extent	Reasonably well	Highly effective program
2	What percentage of design is done with the aid of the computer?	<10%	10–30%	30–65%	65–100%
3	Does the organization extensively use the finite element method for analysis purposes?	Not at all	To some extent	Reasonably well	To a great extent
4	Does the design program include aesthetics and ergonomics of the product?	Not at all	To some extent	Reasonably well	To a great extent
5	Does your organization use simulation and modeling for analyzing designs?	Not at all	Occasionally	Usually	To a great extent

6	Does your organization track design and development program costs?	Not at all	Occasionally	Usually	To a great extent

C – PROCESS PLANNING

1	Does your organization have an effective process planning program?	Not at all	To Some Extent	Reasonably Well	Highly effective Program
2	What percentage of the process planning is done with the aid of the computer?	<10%	10–40%	40–70%	70–100%
3	Does your organization apply group technology while planning?	Not at all	Occasionally	Usually	To a great extent
4	Does your organization possess a mechanism for material and machine selection?	Not at all	Occasionally	Usually	To a great extent
5	Is the planning software updated and reviewed periodically in accordance with technological changes?	Not at all	Occasionally	Usually	To a great extent
6	Does your organization track process planning series costs?	Not at all	Occasionally	Usually	To a great extent

D – RAW MATERIAL AND EQUIPMENT

1	Does your organization use a computer for analysis and record keeping?	Not at all	To some extent	Reasonably well	To a great extent
2	Does your organization use their own transportation?	Never	Sometimes	Mostly	Always

3	Does your organization have enough warehouses for inventory storage?	Not at all	To some extent	Reasonably well	To a great extent
4	Are the three departments (marketing, design, and production) involved in equipment selection decisions?	Not at all	Occasionally	Usually	To a great extent
5	Does your organization have sufficient automated equipment to meet market demands?	Not at all	Occasionally	Usually	To a great extent
E – PRODUCTION PLANNING AND CONTROL					
1	Does your organization use computerized manufacturing systems (CAMs)?	Not at all	Occasionally	Usually	To a great extent
2	How much do you exert to get precise and accurate dimensions?	<30%	30–50%	50–70%	>70%
3	Does your organization prefer green manufacturing?	Not at all	Occasionally	Usually	To a great extent
4	What percentage of the work is done by robots?	<20%	20–40%	40–60%	>60%
5	What percentage of maintenance hours are needed, broken down per day?	<2%	2–10%	10–20%	>20%
6	To what extent are hydraulic and pneumatic systems are employed in your organization?	<20%	20–40%	40–60%	>60%

7	Does your organization track production planning and control program costs?	Not at all	Occasionally	Usually	To a great extent
F – QUALITY CONTROL					
1	Does your organization prefer product testing under actual conditions?	Not at all	To some extent	Reasonably well	Highly effective program
2	Does your organization carry out a life cycle analysis of the product?	Not at all	To some extent	Reasonably well	Highly effective program
3	To what extent does the inspection process in your organization improve the quality?	<20%	20–40%	40–60%	>60%
4	Does your organization use the computer to analyze quality?	Not at all	To some extent	Reasonably well	Highly effective program
5	Does your organization invest in quality control and inspection?	Not at all	To some extent	Reasonably well	Highly effective program
6	How are quality control instructions issued?	Verbally	Handwritten	Pre-typed job cards	Computerized system
G – STRATEGIC AGILITY					
How does competency lead to the following factors for strategy in an organization?		No correlation at all	Nominal effect	Reasonable effect	High correlation
		1	2	3	4
1	Quality conformance	1	2	3	4
2	Improving customer base	1	2	3	4
3	Developing market share	1	2	3	4

4	Achieving higher profit	1	2	3	4
5	Prices of the products	1	2	3	4

H – MANAGEMENT

How does competency achieve the following strategy factors in an organization?	No correlation at all	Nominal effect	Reasonable effect	High correlation
	1	2	3	4
1	Enhanced production capabilities and improved control	1	2	3	4
2	Better planning functions	1	2	3	4
3	Efficient office administration and management	1	2	3	4
4	Policy formation	1	2	3	4
5	Does your management cover risk situations?	Not at all	To some extent	Reasonably well	To a great extent
6	Does your management plan for crisis situations?	Not at all	To some extent	Reasonably well	To a great extent
7	Does proper coordination exist between departments in your organization?	Not at all	To some extent	Reasonably well	To a great extent

I – TEAMWORK

How are the following strategy factors affected by competency in an organization?	No correlation at all	Nominal Effect	Reasonable effect	High correlation
	1	2	3	4
1	Development or fostering of next-generation technology	1	2	3	4

2	Transforming a traditional hierarchical organization into a boundary-less organization	1	2	3	4
3	Promotion of developed product	1	2	3	4
4	Culture of Kaizen and continuous improvement	1	2	3	4
5	Overall equipment effectiveness (OEE) improvement	1	2	3	4
6	Effectively managing process capability (C_{pk})	1	2	3	4
7	Enhanced autonomous maintenance capabilities	1	2	3	4
8	Communication and cooperation among team members	1	2	3	4

J – INTERPERSONAL

How does competency leads to the following factors for strategy in an organization?	No correlation at all	Nominal effect	Reasonable effect	High correlation
	1	2	3	4
1	Safety and health awareness among workers	1	2	3	4
2	Self-confidence of employees	1	2	3	4
3	Stress management	1	2	3	4
4	Waste utilization	1	2	3	4
4	Multi skilled workers	1	2	3	4
5	Broader job perspectives and employee empowerment	1	2	3	4

6	Self-managed project teams and problem-solving groups	1	2	3	4

K – OUTPUT

How is the impact of manufacturing competencies on the following strategic success factors in an organization?

1	Production capacity	Poor	Satisfactory	Good	Excellent
2	Production time	Poor	Satisfactory	Good	Excellent
3	Lead time	Poor	Satisfactory	Good	Excellent
4	Quality	Not at all	To some extent	Reasonably well	To a great extent
5	Reliability	Not at all	To some extent	Reasonably well	To a great extent
6	Productivity	Very low	Low	Effective	Highly effective
7	Growth and expansion	Poor	Satisfactory	Good	Excellent
8	Competitiveness	Not at all	To some extent	Reasonably well	To a great extent
9	Sales (annually)	Very low	Low	Effective	Highly effective
10	Profit (annually)	Very low	Low	Effective	Highly effective
11	Market share	Not at all	To some extent	Reasonably well	To a great extent
12	Customer base	Not at all	To some extent	Reasonably well	To a great extent

(Please tick the appropriate choice)

(SIGNATURE OF RESPONDENT AND SEAL OF THE ORGANIZATION)

Appendix – B

Analytical Hierarchy Process Questionnaire

Product Concept	1	3	5	7	9	Product Design & Development
Product Concept	1	3	5	7	9	Process Planning
Product Concept	1	3	5	7	9	Raw Material & Equipment
Product Concept	1	3	5	7	9	Production Planning & Control
Product Concept	1	3	5	7	9	Quality Control
Product Concept	1	3	5	7	9	Strategic Agility
Product Concept	1	3	5	7	9	Management
Product Concept	1	3	5	7	9	Administration
Product Concept	1	3	5	7	9	Teamwork
Product Concept	1	3	5	7	9	Interpersonal
Product Design & Development	1	3	5	7	9	Process Planning
Product Design & Development	1	3	5	7	9	Raw Material & Equipment
Product Design & Development	1	3	5	7	9	Production Planning & Control
Product Design & Development	1	3	5	7	9	Quality Control
Product Design & Development	1	3	5	7	9	Strategic Agility
Product Design & Development	1	3	5	7	9	Management
Product Design & Development	1	3	5	7	9	Administration
Product Design & Development	1	3	5	7	9	Teamwork
Product Design & Development	1	3	5	7	9	Interpersonal
Process Planning	1	3	5	7	9	Raw Material & Equipment
Process Planning	1	3	5	7	9	Production Planning & Control
Process Planning	1	3	5	7	9	Quality Control
Process Planning	1	3	5	7	9	Strategic Agility
Process Planning	1	3	5	7	9	Management
Process Planning	1	3	5	7	9	Administration
Process Planning	1	3	5	7	9	Teamwork
Process Planning	1	3	5	7	9	Interpersonal

Raw Material & Equipment	1	3	5	7	9	Production Planning & Control
Raw Material & Equipment	1	3	5	7	9	Quality Control
Raw Material & Equipment	1	3	5	7	9	Strategic Agility
Raw Material & Equipment	1	3	5	7	9	Management
Raw Material & Equipment	1	3	5	7	9	Administration
Raw Material & Equipment	1	3	5	7	9	Teamwork
Raw Material & Equipment	1	3	5	7	9	Interpersonal
Production Planning & Control	1	3	5	7	9	Quality Control
Production Planning & Control	1	3	5	7	9	Strategic Agility
Production Planning & Control	1	3	5	7	9	Management
Production Planning & Control	1	3	5	7	9	Administration
Production Planning & Control	1	3	5	7	9	Teamwork
Production Planning & Control	1	3	5	7	9	Interpersonal
Quality Control	1	3	5	7	9	Strategic Agility
Quality Control	1	3	5	7	9	Management
Quality Control	1	3	5	7	9	Administration
Quality Control	1	3	5	7	9	Teamwork
Quality Control	1	3	5	7	9	Interpersonal
Strategic Agility	1	3	5	7	9	Management
Strategic Agility	1	3	5	7	9	Administration
Strategic Agility	1	3	5	7	9	Teamwork
Strategic Agility	1	3	5	7	9	Interpersonal
Management	1	3	5	7	9	Administration
Management	1	3	5	7	9	Teamwork
Management	1	3	5	7	9	Interpersonal
Administration	1	3	5	7	9	Teamwork
Administration	1	3	5	7	9	Interpersonal
Teamwork	1	3	5	7	9	Interpersonal

Appendix – C

Letter of Support

1. HMSI

TO WHOM IT MAY CONCERN

We are pleased to know that Mr. Chandan Deep Singh is pursuing his Ph. D. research work on evaluation of manufacturing competency on strategic success of automobile manufacturing unit. His title of research work "Impact of Manufacturing Competency on Strategic Success of Automobile Industry" is highly related with the work undertaken by us in the last few years, since we are also committed towards enhancing competitiveness.

Manufacturing Competency and Strategic Success model for the Automobile Manufacturing Units is long overdue since in industrial sector, competencies and strategic success are inter-related.

We sincerely feel that the proposed research work involving "Impact of Manufacturing Competency on Strategic Success of Automobile Industry" being undertaken by Mr. Chandan Deep Singh will become a facilitator for automobile Manufacturing Units to adapt the manufacturing competency-strategic success model for achieving objectives.

We wish Mr. Chandan Deep Singh all the best for his endeavors for his research on evaluation of manufacturing competency on strategic success of automobile manufacturing unit, titled "Impact of Manufacturing Competency on Strategic Success of Automobile Industry" and assure to provide him all support for the completion of the research work.

JAGTAR SINGH SOHI
SENIOR EXECUTIVE

2. Suzuki

TO WHOM IT MAY CONCERN

We are pleased to know that Mr. Chandan Deep Singh is pursuing his Ph. D. research work on evaluation of manufacturing competency on strategic success of automobile manufacturing unit. His title of research work "Impact of Manufacturing Competency on Strategic Success of Automobile Industry" is highly related with the work undertaken by us in the last few years, since we are also committed towards enhancing competitiveness.

Manufacturing Competency and Strategic Success model for the Automobile Manufacturing Units is long overdue since in industrial sector, competencies and strategic success are inter-related.

We sincerely feed that the proposed research work involving "Impact of Manufacturing Competency on Strategic Success of Automobile Industry" being undertaken by Mr. Chandan Deep Singh will become a facilitator for automobile Manufacturing Units to adapt the manufacturing competency-strategic success model for achieving objectives.

We wish Mr. Chandan Deep Singh all the best for his endeavors for his research on evaluation of manufacturing competency on strategic success of automobile manufacturing unit, titled "Impact of Manufacturing Competency on Strategic Success of Automobile Industry" and assure to provide him all support for the completion of the research work.

JASPINDER SINGH
DEPUTY MANAGER

3. SML ISUZU

<u>TO WHOM IT MAY CONCERN</u>

We are pleased to know that Mr. Chandan Deep Singh is pursuing his Ph. D. research work on evaluation of manufacturing competency on strategic success of automobile manufacturing unit. His title of research work "Impact of Manufacturing Competency on Strategic Success of Automobile Industry" is highly related with the work undertaken by us in the last few years, since we are also committed towards enhancing competitiveness.

Manufacturing Competency and Strategic Success model for the Automobile Manufacturing Units is long overdue since in industrial sector, competencies and strategic success are inter-related.

We sincerely feed that the proposed research work involving "Impact of Manufacturing Competency on Strategic Success of Automobile Industry" being undertaken by Mr. Chandan Deep Singh will become a facilitator for automobile Manufacturing Units to adapt the manufacturing competency-strategic success model for achieving objectives.

We wish Mr. Chandan Deep Singh all the best for his endeavors for his research on evaluation of manufacturing competency on strategic success of automobile manufacturing unit, titled "Impact of Manufacturing Competency on Strategic Success of Automobile Industry" and assure to provide him all support for the completion of the research work.

Manjinder Singh
20/09/2013
M No
(C09417082836)
Jr. Manager, R&D
SML ISUZU LTD

SANJIV KUMAR SHARMA
MANAGER – R&D

4. Mahindra

TO WHOM IT MAY CONCERN

We are pleased to know that Mr. Chandan Deep Singh is pursuing his Ph. D. research work on evaluation of manufacturing competency on strategic success of automobile manufacturing unit. His title of research work "Impact of Manufacturing Competency on Strategic Success of Automobile Industry" is highly related with the work undertaken by us in the last few years, since we are also committed towards enhancing competitiveness.

Manufacturing Competency and Strategic Success model for the Automobile Manufacturing Units is long overdue since in industrial sector, competencies and strategic success are inter-related.

We sincerely feed that the proposed research work involving "Impact of Manufacturing Competency on Strategic Success of Automobile Industry" being undertaken by Mr. Chandan Deep Singh will become a facilitator for automobile Manufacturing Units to adapt the manufacturing competency-strategic success model for achieving objectives.

We wish Mr. Chandan Deep Singh all the best for his endeavors for his research on evaluation of manufacturing competency on strategic success of automobile manufacturing unit, titled "Impact of Manufacturing Competency on Strategic Success of Automobile Industry" and assure to provide him all support for the completion of the research work.

DINESH SHARMA
DEPUTY MANAGER – PLANT 2
Heavy Machine shop, Plant-2
Mahindra & mahindra Swaraj

References

Ahmad, S. and Schroeder, R. G. (2011), "Knowledge management through technology strategy: Implications for competitiveness", *Journal of Manufacturing Technology Management*, Vol. 22, No. 1, pp. 6–24.

Alsudiri, T., Al-Karaghouli, W. and Eldabi, T. (2013), "Alignment of large project management process to business strategy: A review and conceptual framework", *Journal of Enterprise Information Management*, Vol. 26, No. 5, pp. 596–615.

Amoako-Gyampah, K. and Acquaah, M. (2008), "Manufacturing strategy, competitive strategy and firm performance: An empirical study in a developing economy environment", *International Journal of Production Economics*, Vol. 111, No. 2, pp. 575–592.

Amoako-Gyampah, K., Acquaah, M. and Jayaram, J. M. (2007), "The effects of human resource management, manufacturing and marketing strategies on competitive strategy and firm performance in an emerging economy", *39th Annual Meeting*, Decision Sciences Institute, pp. 502–503.

Armstrong, C. E. (2013), "Competence or flexibility? Survival and growth implications of competitive strategy preferences among small US businesses", *Journal of Strategy and Management*, Vol. 6, No. 4, pp. 377–398.

Averyt, W. F. and Ramagopal, K. (1999), "Strategic disruption and transaction cost economics: The case of the American auto industry and Japanese competition", *International Business Review*, Vol. 8, No. 1, pp. 39–53.

Barrett, P. (2007), "Structural equation modeling: Adjudging model fit", *Personality and Individual Differences*, Vol. 42, No. 3, pp. 815–824.

Belkadi, F., Bonjour, E. and Dulmet, M. (2007), "Competency characterisation by means of work situation modeling", *The Journal Computers in Industry*, Vol. 58, No. 2, pp. 85–91.

Bhamra, R., Dani, S. and Bhamra, T. (2010), "Competence understanding and use in SMEs: A UK manufacturing perspective", *International Journal of Production Research*, Vol. 49, pp. 2729–2743.

Bonjour, E. and Micaelli, J.-P. (2010), "Design core competence diagnosis: A case from the automotive industry", *IEEE Transactions on Engineering Management*, Vol. 57, No. 2, pp. 323–337.

Byrne, B. M. (1989), *A Primer of LISREL: Basic Applications and Programming for Confirmatory Factor Analytic Models*, Springer-Verlag, New York, ISBN: 978-1-4613-8887-6.

Cardy, R. L. and Selvarajan, T. T. (2006), "Competencies: Alternative frameworks for competitive advantage", *Business Horizons*, Vol. 49, No. 3, pp. 235–245.

Chaiprasit, S. and Swierczek, F. W. (2011), "Competitiveness, globalization and technology development in Thai firms", *Competitiveness Review: An International Business Journal*, Vol. 21, No. 2, pp. 188–204.

Chang, H.-C., Lai, -H.-H. and Chang, Y.-M. (2005), "Expression modes used by consumers in conveying desire for product form: A case study of a car", *International Journal of Industrial Ergonomics*, Vol. 36, No. 1, pp. 3–10.

Chen, C.-J., Hsiao, Y.-C. and Chu, M.-A. (2014), "Transfer mechanisms and knowledge transfer: The cooperative competency perspective", *Journal of Business Research*, Vol. 67, No. 12, pp. 2531–2541.

Coyne, K. P. (1986), "Sustainable competitive advantage: What it is, what it isn't", *Business Horizons*, Vol. 29, No. 1, pp. 54–61.

Crema, M., Verbano, C. and Venturini, K. (2014), "Linking strategy with open innovation and performance in SMEs", *Measuring Business Excellence*, Vol. 18, No. 2, pp. 14–27.

Currie, D. J., Francis, A. P. and Kerr, J. T. (1999), "Some general propositions about the study of spatial patterns of species richness", *Ecoscience*, Vol. 6, pp. 392–399.

Dangayach, G. S. and Deshmukh, S. G. (2000), "Manufacturing strategy: Experiences from select Indian organizations", *Journal of Manufacturing Systems*, Vol. 19, No. 2, pp. 134–148.

Dangayach, G. S., Pathak, S. C. and Sharma, A. D. (2006), "Advanced manufacturing technology: A way of improving technological competitiveness", *International Journal of Global Business and Competitiveness*, Vol. 2, No. 1, pp. 1–8.

Demirtas, O. (2013), "Evaluating the core capabilities for strategic outsourcing decisions at aviation maintenance industry", *Procedia – Social and Behavioral Sciences*, Vol. 99, pp. 1134–1143.

Drauz, R. (2014), "Re-insourcing as a manufacturing-strategic option during a crisis-cases from the automobile industry", *Journal of Business Research*, Vol. 67, No. 3, pp. 346–353.

Drejer, A. (2001), "How can we define and understand competencies and their development?", *The Journal Technovation*, Vol. 21, pp. 135–146.

Dutta, S. K. (2007), "Enhancing competitiveness of India Inc.", *International Journal of Social Economics*, Vol. 34, No. 9, pp. 679–711.

Fernández-Mesa, A., Alegre-Vidal, J., Chiva-Gómez, R. and Gutiérrez-Gracia, A. (2013), "Design management capability and product innovation in SMEs", *Management Decision*, Vol. 51, No. 3, pp. 547–565.

Fernández-Mesa, A., Ferreras-Méndez, J. L., Alegre, J. and Chiva, R. (2014), "IT competency and the commercial success of innovation", *Industrial Management and Data Systems*, Vol. 114, No. 4, pp. 550–567.

Fernández-Pérez, V., García-Morales, V. J., Fernando, O. and SáNchez, B. (2012), "The effects of CEOs' social networks on organizational performance through knowledge and strategic flexibility", *Personnel Review*, Vol. 41, No. 6, pp. 777–812.

Flint, R. (2000), *Functional Competencies and Their Effects on Performance of Manufacturing Companies in Vietnam*, PhD Thesis, University of Fribourg, Switzerland.

Fleury, A. and Fleury, M. T. (2003), "Competitive strategies and core competencies: Perspectives for the internationalization of industry in Brazil", *Integrated Manufacturing Systems*, Vol. 14, No. 1, pp. 16–25.

Fredriksson, P. (2004), "Modular assembly in the car industry-an analysis of organizational influence on performance", *European Journal of Purchasing and Supply Management*, Vol. 8, No. 4, pp. 221–233.

Gallon, M. R., Stillman, H. M. and Coates, D. (1999), "Putting core competency thinking into practice", *Research Technology Management Journal*, Vol. 38, No. 3, pp. 20–28.

Geels, F. W. (2001), "Technological transitions as evolutionary reconfiguration processes: A multi-level perspective and a case-study", *Research Policy*, Vol. 31, pp. 1257–1274.

Haartman, R. V. (2012), "Manufacturing capabilities: Mere drivers of operational performance or critical for customer-driven innovation", *Proceedings of 4th Joint World Conference on Production and Operations Management*, pp. 1–10.

Hair, J. F., Black, W. C., Babin, B. J. and Anderson, R. E. (2005), *Multivariate Data Analysis*, (7th ed.), Pearson Prentice Hall, Upper Saddle River, NJ, ISBN: 978-1-2920-2190-4.

Hanzaee, K. H. and Sadeghian, M. (2014), "The impact of corporate social responsibility on customer satisfaction and corporate reputation in automotive industry: Evidence from Iran", *Journal of Islamic Marketing*, Vol. 5, No. 1, pp. 125–143.

Hong, J. and Ståhle, P. (2005), "The coevolution of knowledge and competence management", *International Journal of Management Concepts and Philosophy*, Vol. 1, No. 2, pp. 129–145.

Janssen, W., Bouwman, H., Buuren, R. V. and Haaker, T. (2014), "An organizational competence model for innovation intermediaries", *European Journal of Innovation Management*, Vol. 17, No. 1, pp. 2–24.

Jeong, K. and Phillips, D. T. (2001), "Operational efficiency and effectiveness measurement", *International Journal of Operations and Production Management*, Vol. 21, No. 11, pp. 1404–1416.

Kassahun, A. E. and Molla, A. (2013), "BPR complementary competence: Definition, model and measurement", *Business Process Management Journal*, Vol. 19, No. 3, pp. 575–596.

Khanna, V. K. and Gupta, R. (2014), "Comparative study of the impact of competency-based training on 5 'S' and TQM: A case study", *International Journal of Quality and Reliability Management*, Vol. 31, No. 3, pp. 238–260.

Laosirihongthong, T. and Dangayach, G. S. (2005), "A comparative study of implementation of manufacturing strategies in Thai and Indian automotive manufacturing companies", *Journal of Manufacturing Systems*, Vol. 24, No. 2, pp. 131–143.

Lee, Y.-T. (2006), "Exploring high-performers' required competencies", *Expert Systems with Applications*, Vol. 37, No. 1, pp. 434–439.

Levinthal, D. A. and March, J. G. (1993), "The myopia of learning", *Strategic Management Journal*, Vol. 14, No. S2, pp. 95–112.

Lim, J.-S., Sharkey, T. W. and Heinrichs, J. H. (2006), "Strategic impact of new product development on export involvement", *European Journal of Marketing*, Vol. 40, No. 1/2, pp. 44–60.

Linton, J. and Walsh, S. (2013), "The effect of technology on learning during the acquisition and development of competencies in technology-intensive small firms", *International Journal of Entrepreneurial Behaviour and Research*, Vol. 19, No. 2, pp. 165–186.

Ljungquist, U. (2007), "Core competency beyond identification: Presentation of a model", *Management Decision*, Vol. 45, No. 3, pp. 393–408.

Ljungquist, U. (2013), "Adding dynamics to core competence concept applications", *European Business Review*, Vol. 25, No. 5, pp. 453–465.

Maran, M., Thiagarajan, K., Manikandan, G. and Sarukesi, K. (2016), "Competency enhancement and employee empowerment in a TPM organization – An empirical study", *International Journal of Advanced Engineering Technology*, Vol. 7, No. 2, pp. 40–47.

Masoud, E. Y. (2013), "The impact of functional competencies on firm performance of pharmaceutical industry in Jordan", *International Journal of Marketing Studies*, Vol. 5, No. 3, pp. 56–72.

Meybodi, M. Z. (2013), "The links between lean manufacturing practices and concurrent engineering method of new product development: An empirical study", *Benchmarking: An International Journal*, Vol. 20, No. 3, pp. 362–376.

Millikin, J. P., Hom, P. W. and Manz, C. C. (2010), "Self-management competencies in self-managing teams: Their impact on multi-team system productivity", *Leadership Quarterly*, Vol. 21, pp. 687–702.

Mishra, R., Pundir, A. K. and Ganapathy, L. (2014), "Assessment of manufacturing flexibility: A review of research and conceptual framework", *Management Research Review*, Vol. 37, No. 8, pp. 750–776.

Mitchelmore, S. and Rowley, J. (2010), "Entrepreneurial competencies: A literature review and development agenda", *International Journal of Entrepreneurial Behaviour and Research*, Vol. 16, No. 2, pp. 92–111.

Moom, B.-J. (2013), "Antecedants and outcomes of strategic thinking", *Journal of Business Research*, Vol. 66, No. 10, pp. 1698–1708.

Moore, F. (2011), "Identity, knowledge and strategy in the UK subsidiary of an Anglo-German automobile manufacturer", *International Business Review*, Vol. 21, No. 2, pp. 281–292.

Mothe, C. and Quelin, B. (2000), "Creating competencies through collaboration. The case of Eureka R and D Consortia", *European Management Journal*, Vol. 18, No. 6, pp. 590–604.

Muffatto, M. (1998), "Reorganizing for product development: Evidence from Japanese automobile firms", *International Journal of Production Economics*, Vol. 56–57, pp. 483–493.

Nicolin, C. (1983), "The power of competency in industrial development", *Technology in Society*, Vol. 5, No. 3–4, pp. 167–170.

Nunnaly, J. C. (1978), *Psychometric Theory*, McGraw-Hill, New York, ISBN: 978-0-07-107088-1.

Nuntamanop, P., Kauranen, I. and Igel, B. (2013), "A new model of strategic thinking competency", *Journal of Strategy and Management*, Vol. 6, No. 3, pp. 242–264.

Nyhan, B. (1998), "Competence development as a key organisational strategy – Experiences of European companies", *Industrial and Commercial Training*, Vol. 30, No. 7, pp. 267–273.

Omar, N. M. and Fayek, R. A. (2016), "Modeling and evaluating construction project competencies and their relationship to project performance", *Automation in Construction*, Vol. 69, pp. 115–130.

Pallant, J. (2005), *SPSS Survival Manual: A Step by Step Guide to Data Analysis Using SPSS for Window*, Allen and Unwin, New South Wales, Australia.

Park, T. and Rhee, J. (2012), "Antecedents of knowledge competency and performance in born globals, the moderating effects of absorptive capacity", *Management Decision*, Vol. 50, No. 8, pp. 1361–1381.

Patel, P. and Pavitt, K. (1997), "The technological competencies of the world's largest firms: Complex and path- dependent, but, not much variety", *Research Policy*, Vol. 26, No. 2, pp. 141–156.

Pham, D. T. and Thomas, A. J. (2012), "Fit manufacturing: A framework for sustainability", *Journal of Manufacturing Technology Management*, Vol. 23, No. 1, pp. 103–123.

Porter, M. E. (1996), "Changing patterns of international competition", *California Management Review*, Vol. 28, No. 2, pp. 9–40.

Potnuru, R. K. G. and Sahoo, C. K. (2016), "HRD interventions, employee competencies and organizational effectiveness: An empirical study", *European Journal of Training and Development*, Vol. 40, No. 5, pp. 345–365.

Prahalad, E. K. and Hamel, G. (1994), "The core competence of the corporation", *Harvard Business Review*, Vol. 68, No. 4, pp. 79–93.

Rakowski, W., Andersen, M. R. and Stoddard, A. M. (1997), "Confirmatory analysis of opinions regarding the pros and cons of mammography", *Health Psychology*, Vol. 16, No. 2, pp. 433–444.

Reijnders, L., Loureiro, S. M. C. and Sardinha, I. M. D. (2012), "The effect of corporate social responsibility on consumer satisfaction and perceived value: The case of the automobile industry sector in Portugal", *Journal of Cleaner Production*, Vol. 37, pp. 172–178.

Ryan, G., Spencer, L. M. and Bernhard, U. (2012), "Development and validation of a customized competency-based questionnaire linking social, emotional and cognitive competencies to business unit profitability", *Cross Cultural Management*, Vol. 19, No. 1, pp. 90–103.

Saaty, T. L. (1980), *The Analytic Hierarchy Process*, McGraw-Hill Book Co., New York, ISBN: 978-0-9620-3172-4.

Saaty, T. L. (1994), *Fundamentals of Decision Making*, RWS Publications, Pittsburgh, PA, ISBN: 1888603151.

Sahoo, T., Banwet, D. K. and Momaya, K. (2011), "Strategic technology management in the auto component industry in India A case study of select organizations", *Journal of Advances in Management Research*, Vol. 8, No. 1, pp. 9–29.

Schlie, E. and Yip, G. (2000), "Regional follows global: Strategy mixes in the world automotive industry", *European Management Journal*, Vol. 18, pp. 234–240.

Sengupta, A., Venkatesh, D. N. and Sinha, A. K. (2013), "Developing performance-linked competency model: A tool for competitive advantage", *International Journal of Organizational Analysis*, Vol. 21, No. 4, pp. 504–527.

Sharma, M. and Kodali, R. (2008), "Development of a framework for manufacturing excellence", *Measuring Business Excellence*, Vol. 12, No. 4, pp. 50–66.

Shavarini, S. K., Salimian, H., Nazemi, J. and Alborzi, M. (2013), "Operations strategy and business strategy alignment model (case of Iranian industries)", *International Journal of Operations and Production Management*, Vol. 33, No. 9, pp. 1108–1130.

Shinnaranantana, N., Dimmitt, N. J. and Siengthai, S. (2013), "CSR manager competencies: A case study from Thailand", *Social Responsibility Journal*, Vol. 9, No. 3, pp. 395–411.

Singh, H. and Mahmood, R. (2014), "Aligning manufacturing strategy to export performance of manufacturing small and medium enterprises in Malaysia", *Procedia-Social and Behavioral Sciences*, Vol. 130, pp. 85–95.

Singh, K. and Ahuja, I. S. (2012), "Justification of TQM–TPM implementations in manufacturing organisations using analytical hierarchy process: A decision-making approach under uncertainty", *International Journal of Productivity and Quality Management*, Vol. 10, No. 1, pp. 69–84.

Singh, R. K., Garg, S. K. and Deshmukh, S. G. (2007), "Strategy development for competitiveness: A study on Indian auto component sector", *International Journal of Productivity and Performance Management*, Vol. 56, No. 4, pp. 285–304.

Singh, R. K., Garg, S. K. and Deshmukh, S. G. (2008), "Competency and performance analysis of Indian SMEs and large organizations an exploratory study", *Competitiveness Review: An International Business Journal*, Vol. 18, No. 4, pp. 308–320.

Singh, R. K., Garg, S. K. and Deshmukh, S. G. (2010), "Strategy development by small scale industries in India", *Industrial Management and Data Systems*, Vol. 110, No. 7, pp. 1073–1093.

Sleuwaegen, L. (2013), "Scanning for profitable (international) growth", *Journal of Strategy and Management*, Vol. 6, No. 1, pp. 96–110.

Soderquist, K. E., Papalexandris, A., Ioannou, G. and Prastacos, G. (2010), "From task-based to competency-based a typology and process supporting a critical HRM transition", *Personnel Review*, Vol. 39, No. 3, pp. 325–346.

Srinivasan, V. and Shekhar, B. (2000), "Application of the uncertainty-based mental model in organizational learning", *Accounting, Management and Information Technologies*, Vol. 7, No. 2, pp. 87–112.

Steptoe-Warren, G., Howat, D. and Hume, I. (2011), "Strategic thinking and decision making: Literature review", *Journal of Strategy and Management*, Vol. 4, No. 3, pp. 238–250.

Stokes, P. and Oiry, E. (2012), "An evaluation of the use of competencies in human resource development – A historical and contemporary recontextualisation", *Euro-Med Journal of Business*, Vol. 7, No. 1, pp. 4–23.

Subramoniam, R., Huisingh, D. and Chinnam, R. B. (2009), "Remanufacturing for the automotive aftermarket-strategic factors: Literature review and future research needs", *The Journal of Cleaner Production*, Vol. 17, pp. 172–181.

Sutton, A. and Watson, S. (2013), "Can competencies at selection predict performance and development needs?", *Journal of Management Development*, Vol. 32, No. 9, pp. 1023–1035.

Tarafdar, M. and Gordon, S. R. (2009), "Understanding the influence of information systems competencies on process innovation", *The Journal of Strategic Information Systems*, Vol. 16, No. 4, pp. 357–363.

Teece, D., Pisano, G. and Shuen, A. (1997), "Dynamic capabilities and strategic management", *Strategic Management Journal*, Vol. 18, No. 7, pp. 509–534.

Thomas, A. J., Byard, P. and Evans, R. (2012), "Identifying the UK's manufacturing challenges as a benchmark for future growth", *Journal of Manufacturing Technology Management*, Vol. 23, No. 2, pp. 142–156.

Ullman, J. B. (2001), *Structural equation modeling* In B. G. Tabachnick and L. S. Fidell (eds.), *Using Multivariate Statistics*, (4th ed.), Allyn and Bacon, Needham Heights, MA, pp. 653–771.

Verle, K., Markic, M., Kodric, B. and Zoran, A. G. (2014), "Managerial competencies and organizational structures", *Industrial Management and Data Systems*, Vol. 114, No. 6, pp. 922–935.

Waal, A. and Kourtit, K. (2013), "Performance measurement and management in practice: Advantages, disadvantages and reasons for use", *International Journal of Productivity and Performance Management*, Vol. 62, No. 5, pp. 446–473.

Wang, K.-J. and Lestari, Y. D. (2013), "Firm competencies on market entry success: Evidence from a high-tech industry in an emerging market", *Journal of Business Research*, Vol. 66, No. 12, pp. 2444–2450.

Wang, Y., Hing-Po, L. and Yang, Y. (2004), "The constitutes of the core competencies and firm performance: Evidence from high-technology firms in China", *Journal of Engineering and Technology Management*, Vol. 21, No. 4, pp. 249–280.

Wee, B. V., Sierzchula, W., Bakker, S. and Maat, K. (2012), "Technological diversity of emerging eco-innovations: A case study of the automobile industry", *Journal of Cleaner Production*, Vol. 37, pp. 211–220.

Williams, A. (2006), "Product service systems in the automobile industry: Contribution to system innovation", *Journal of Cleaner Production*, Vol. 15, No. 11–12, pp. 1093–1103.

Yazdanfar, D., Abbasian, S. and Hellgren, C. (2014), "Competence development and performance among Swedish micro firms", *European Journal of Training and Development*, Vol. 38, No. 3, pp. 162–179.

Yildirim, Y. Y., Akar, C., Akar, S. and Celik, C. (2012), "The relationship between automobile loans and automobile production amount as a key factor for production strategy", *Procedia – Social and Behavioral Sciences*, Vol. 58, pp. 1476–1481.

Zhang, Q., Vonderembse, M. A. and Lim, J.-S. (2003), "Manufacturing flexibility: Defining and analyzing relationships among competence, capability and customer satisfaction", *Journal of Operations Management*, Vol. 21, pp. 540–547.

Zhi-Yu, W., Yan-Lin, Q. and Shi-He, G. (2006), "Quality competence: A source of sustained competitive advantage", *The Journal of China Universities of Posts and Telecommunications*, Vol. 13, No. 1, pp. 104–108.

Web References

http://en.wikipedia.org/wiki/competence_human
www.carltonglobal.com/samplelesson_competency
www.competencyworks.org
www.careeronstop.org/competency model
www.astd.org/competency model
www.ask.com/defination of competency
www.studymode.com

Index

Milton Keynes UK
Ingram Content Group UK Ltd.
UKHW040103071024
449327UK00019B/774